Springer Theses

Recognizing Outstanding Ph.D. Research

Aims and Scope

The series "Springer Theses" brings together a selection of the very best Ph.D. theses from around the world and across the physical sciences. Nominated and endorsed by two recognized specialists, each published volume has been selected for its scientific excellence and the high impact of its contents for the pertinent field of research. For greater accessibility to non-specialists, the published versions include an extended introduction, as well as a foreword by the student's supervisor explaining the special relevance of the work for the field. As a whole, the series will provide a valuable resource both for newcomers to the research fields described, and for other scientists seeking detailed background information on special questions. Finally, it provides an accredited documentation of the valuable contributions made by today's younger generation of scientists.

Theses are accepted into the series by invited nomination only and must fulfill all of the following criteria

- They must be written in good English.
- The topic should fall within the confines of Chemistry, Physics, Earth Sciences, Engineering and related interdisciplinary fields such as Materials, Nanoscience, Chemical Engineering, Complex Systems and Biophysics.
- The work reported in the thesis must represent a significant scientific advance.
- If the thesis includes previously published material, permission to reproduce this must be gained from the respective copyright holder.
- They must have been examined and passed during the 12 months prior to nomination.
- Each thesis should include a foreword by the supervisor outlining the significance of its content.
- The theses should have a clearly defined structure including an introduction accessible to scientists not expert in that particular field.

More information about this series at http://www.springer.com/series/8790

Luca Fedeli

High Field Plasmonics

Doctoral Thesis accepted by
University of Pisa, Italy

 Springer

Author
Dr. Luca Fedeli
Enrico Fermi Department of Physics
University of Pisa
Pisa
Italy

Supervisor
Dr. Andrea Macchi
"Adriano Gozzini" Unit
CNR National Institute of Optics
Pisa
Italy

ISSN 2190-5053   ISSN 2190-5061 (electronic)
Springer Theses
ISBN 978-3-319-44289-1   ISBN 978-3-319-44290-7 (eBook)
DOI 10.1007/978-3-319-44290-7

Library of Congress Control Number: 2016948806

© Springer International Publishing AG 2017
This work is subject to copyright. All rights are reserved by the Publisher, whether the whole or part of the material is concerned, specifically the rights of translation, reprinting, reuse of illustrations, recitation, broadcasting, reproduction on microfilms or in any other physical way, and transmission or information storage and retrieval, electronic adaptation, computer software, or by similar or dissimilar methodology now known or hereafter developed.
The use of general descriptive names, registered names, trademarks, service marks, etc. in this publication does not imply, even in the absence of a specific statement, that such names are exempt from the relevant protective laws and regulations and therefore free for general use.
The publisher, the authors and the editors are safe to assume that the advice and information in this book are believed to be true and accurate at the date of publication. Neither the publisher nor the authors or the editors give a warranty, express or implied, with respect to the material contained herein or for any errors or omissions that may have been made.

Printed on acid-free paper

This Springer imprint is published by Springer Nature
The registered company is Springer International Publishing AG
The registered company address is: Gewerbestrasse 11, 6330 Cham, Switzerland

To my grandparents

Supervisor's Foreword

Plasmons are collective oscillations of electrons which can be excited either in the bulk of a material or at a sharp interface between two different media. Surface plasmons can be either localized on a small size structure or propagating along an interface. Depending on their wavelength, the surface plasmons can be predominantly of electrostatic or electromagnetic nature (in the latter case, they are also commonly referred to as surface waves), with the field localized in a narrow region across the interface. The excitation of plasmons can be exploited to tailor the interaction of laser light with a material, the simplest example being resonant absorption due to linear conversion of the electromagnetic wave (the laser pulse) into a bulk plasmon. The coupling with surface plasmons in different metals and for different structuring has stimulated the development of several applications based on local enhancement and manipulation of electromagnetic fields, giving birth to plasmonics.

Extending plasmonics schemes to the highest laser intensities available today is of great interest, since enhancement and control of laser-matter coupling in this regime may allow substantial progress in, e.g., the development of secondary sources of high-energy radiation (electrons, ions, and photons). Ultimately, being able to focus and localize laser light in extremely small volumes may allow the study of QED in a novel regime of high fields and high energy densities. At high irradiation intensities, every material (including dielectrics) is ionized almost instantaneously and too a large degree, making free electrons available for plasmon excitation. However, in a solid material the electron density is high enough to make the plasma frequency typically one order of magnitude higher than the laser frequency, making the resonant coupling with bulk plasmons difficult. In addition, the laser-produced plasma is hot enough to expand quickly, so that any structuring designed to allow the coupling with surface plasmons may be washed away in a short time. Thus, the high-field interaction with targets structured at a scale close to the laser wavelength, as needed for coupling to surface plasmons, requires ultrashort (tens of femtosecond) pulses. In addition, the intense femtosecond laser pulse must be free from short prepulses and long pedestals, typical of high-power laser systems, which may be already intense enough to destroy the surface structuring before the main interaction; such issue has prevented the use of structured targets at high intensities for a long time. Beyond these technical hurdles, there is little knowledge about the properties of surface plasmons at field amplitudes high enough to make the electron dynamics strongly nonlinear and relativistic.

The work presented in the thesis by Luca Fedeli is focused on two experiments where recent developments in both intense laser manipulation and material science are exploited for pioneering studies of high-field plasmonics. The first experiment uses a laser system with an exceptionally high contrast, so that a shallow submicrometer graving of a target can survive up to the interaction at very high intensities, in a regime of strongly relativistic electrons. This allowed to couple the laser light with a periodically engraved target, i.e., a grating, which allows the excitation of surface plasmons. The experiment has confirmed that surface plasmons, having a longitudinal field component and a phase velocity slightly lower than that of the speed of light, are able to accelerate electrons along the target surface. Besides the possible application as an intense, ultrashort source of multi-MeV electrons, this study gives the most direct evidence of relativistic surface plasmons so far.

The second experiment is based on the use of a special target material, i.e., a carbon foam with a reduced spatially averaged density, low enough to make the plasma frequency match the laser frequency. Previous theoretical studies showed efficient laser absorption and conversion into high energetic electrons, which in turn leads to enhancement of proton acceleration at the rear side of the target. Albeit the experimental results suggest that further optimization is needed, the observed increase in proton energy with respect to standard targets (uncovered with foams) is already substantial, making the approach promising for ion acceleration and deserving further investigations.

For both experiments, a very remarkable feature is the close agreement of the experimental results with three-dimensional particle-in-cell (PIC) simulations. The validation of the PICcante code used for such simulations makes it a powerful tool for the design of further experimental investigations of high-field plasmonics. The last chapter of the thesis presents a preliminary survey of some of such ideas, mostly related to laser field manipulation and conversion into higher frequency radiation. It is remarkable as well, and a key point in the above-mentioned success of the numerical simulations, that Luca Fedeli has been a main promoter and developer of the PICcante code, showing an impressive ability in computational physics as well. As a noticeable addition, PICcante has been released as an open source project, freely and fully available for the scientific community, with a commitment to make the code flexible for adaption to different geometries and easy to use. A brief presentation of side projects where PICcante has been used is also shown in the thesis.

This thesis is so far the most detailed presentation of experimental and theoretical research in high-field plasmonics. Hopefully, it will be precious for scientist willing to investigate this widely unexplored field, in which Dr. Luca Fedeli is of course expected to make further important contributions.

Pisa, Italy
July 2016

Dr. Andrea Macchi

Abstract

The manuscript concerns the study of plasmonic effects at high fields, that is, in the framework of laser–plasma interaction at ultra-high intensities ($I > 10^{18}$ W/cm^2). "Plasmonics" is the study and application of surface plasmons, which are electromagnetic modes at the interface between a metal and a dielectric medium. Surface plasmons are normally excited with low-intensity laser pulses. This regime is well known from the theoretical point of view and the study of plasmonic schemes is a vibrant research field. On the other hand, a complete theoretical model is still lacking for surface plasmons in the high-intensity regime, where strong nonlinear and relativistic effects might play a relevant role. Also the numerical and experimental investigation of plasmonics in this regime has been limited up to now.

This thesis presents numerical and experimental results related to the study of plasmonic effects at high fields in different scenarios.

The interaction of intense laser pulses ($I \approx 5 \times 10^{19}$ W/cm^2) with microstructured grating targets was studied experimentally at CEA-Saclay (Gif-sur-Yvette, France). We observed the acceleration of collimated electron bunches along the surface of grating target when irradiated at angles close to that expected for the excitation of a surface plasmon. We measured the energy spectrum of the emitted electrons, observing a peaked distribution (around ~ 5 MeV) and total charges greater than 100 pC. These characteristics make the source interesting for some applications (such as ultra-fast electron diffraction). Simulations proved to be in very good agreement with the experimental results. A theoretical model is provided to clarify the role played by surface plasmons in the acceleration process.

The interaction of intense laser pulses ($I > 10^{20}$ W/cm^2) with solid targets coupled with a carbon foam was studied (the activity was carried out at GIST, Gwangju, Republic of Korea). The average density of the carbon foam was selected in order to obtain a plasma at the critical density. Laser–plasma interaction in this regime allows for an efficient coupling of the laser pulse with bulk plasmons, enhancing the efficiency of the energy absorption by the target. The activity was carried out in the framework of ion acceleration with laser-produced plasmas. During the two experimental campaigns, we observed that targets coated with a thin foam allowed to obtain higher ion energies with respect to simple targets. Numerical simulations with particle-in-cell codes helped to clarify the role of the microstructuring of the foam.

We studied the role of plasmonic effects in laser-driven Rayleigh–Taylor instability, which may develop in radiation pressure acceleration scenarios, where thin solid foils are directly accelerated by the radiation pressure of an ultra-intense laser pulse. Using a simple model, it is shown that the self-consistent modulation of the radiation pressure caused by a sinusoidal rippling affects substantially the wavevector spectrum of the instability. The nonlinear evolution is investigated by 3D simulations which show the formation of net-like structures. Other physical scenarios which might involve high-field plasmonics effects (e.g., high-order harmonic emission from irradiated grating targets) are discussed at the end of the manuscript, suggesting possible future works.

Acknowledgments

My Ph.D. years have flown quickly and I've truly enjoyed this journey. Unfortunately, these few acknowledgements lines would inevitably be an insufficient tribute to all the extraordinary people I've met and all the interesting experiences I've accumulated.

First of all my heartfelt thanks go my advisor, Dr. Andrea Macchi. I've learnt a lot under his supervision and he has always been available for a smart advice, a chat or some occasional bitter comment on human nature. He has provided me several interesting research opportunities and I've also greatly appreciated the freedom I've enjoyed in pursuing my own research interests. I would also like to warmly thank my internal supervisor, Prof. Francesco Pegoraro, for his advices. I am grateful to Dr. Andrea Sgattoni for his constant support during my Ph.D. He has taught me everything I know on Particle-In-Cell codes and working with him to write a massively parallel code from scratch was an incredibly formative experience. He has also found the time to proof-read this manuscript a few days before his relocation in Paris. I would like to express my gratitude also to Prof. Fulvio Cornolti, thanks to whom I now know much more fluid dynamics than before my Ph.D.

I owe a lot to my colleagues in Dr. Macchi's group. My Ph.d. would not have been the same without Andrea, Giada, Marta, Anna and Giannandrea.

There are several colleagues and groups I would like to acknowledge here.

I would like to thank Dr. T. Ceccotti and the group at CEA-Saclay (Fabrice, David...) in Paris for the wonderful atmosphere in the lab.

I am indebted to Prof. M. Borghesi (University of Belfast) for allowing me to take part to an experiment at Rutherford Appleton Laboratory (UK).

I would like to thank Prof. M. Passoni (Politencnico di Milano) and his group (David and Irene) for the two experiments in South Korea.

I am grateful to Prof. I.W. Choi (GIST, South Korea) and his colleagues for their nice hospitality. I would like to especially thank Prof. Choi for his passionate introduction to Korean culture and food.

I would like to thank Stefano Sinigardi (University of Bologna) for his precious assistance in the writing of piccante code.

I am grateful to Dr. G. Sarri (University of Belfast) and Dr. M. Tamburini for insightful conversations during conferences.

I am indebted to Prof. A. Morace for inviting me for a seminar at University of Osaka and for the great time we spent in Osaka with him and his wife Yuko.

The city of Pisa has contributed significantly to the quality of my Ph.D. I've appreciated its vibrant life but also its small scale, which allows for a relaxed, relatively slow-paced existence. I've also loved Tuscany in general: history, landscape, food and especially the people. I'll never be baffled enough by local rivalries, which extend up to any scale (from city against city down to district against district). In the fierce rivalry between Pisa and Livorno, I would definitively take the side of Pisa.

I would like to thank my (former and present) flatmates (a special mention goes to Roberto) I am also grateful to my friends in Pisa (Alessandra, Giorgia, Mario, Chiara, Matteo, Luca, Ilaria, Cristiana, Simone, Lisa, Daniele, Guido) for the great time we spent together.

I would like to thank my martial arts instructor, Cristian, who has passionately taught me some Jeet-Kune-Do, and the people I met at the gym.

I would like to thank also a few old friends, with whom I managed to keep in touch: Luca N. and Luca F., Federico, Stefania, Selena, Flavio, Chiara, Simone, Teo, Cristina, Matteo T. and Matteo R.

I am grateful to my brother Davide-Duzzo- and my parents and relatives for their constant support during these years.

Finally, I want to deeply thank my girlfriend Lucia.

Contents

1	**Introduction**		1
	References		5
2	**Introduction on High Intensity Laser-Plasma Interaction and High Field Plasmonics**		7
	2.1	Evolution of High Intensity Laser Technology	7
		2.1.1 Overview	8
		2.1.2 A Typical High Intensity Ti:Sapphire Laser System	10
		2.1.3 Towards 10 PW Laser Systems	11
	2.2	Relativistic Laser Plasma Interaction	12
		2.2.1 Single Particle Motion	12
		2.2.2 Propagation of EM Waves in a Plasma	14
		2.2.3 Relativistic Kinetic Equations	16
		2.2.4 Energy Absorption with Overdense Targets	18
		2.2.5 Target Normal Sheath Acceleration (TNSA)	21
		2.2.6 Radiation Reaction Force and QED Effects	22
		2.2.7 Applications	23
	2.3	High Field Plasmonics	28
		2.3.1 Excitation of Surface Plasmons	29
		2.3.2 Overview of Plasmonic Schemes and Applications	33
		2.3.3 Outlook for Relativistic Plasmonics	34
	References		35
3	**Numerical Tools**		41
	3.1	Numerical Simulations of Plasma Physics	41
	3.2	Particle-In-Cell Codes	43
	3.3	PICCANTE: An Open-Source PIC Code	45
		3.3.1 Optimization of Piccante	48
	3.4	PICcolino: A Spectral PIC Code	52

	3.5	Applications	54
		3.5.1 Weibel Instability in Pair-Plasmas	54
		3.5.2 Intense Laser Interaction with Thin Gold Targets	58
	References		60

4 Electron Acceleration with Grating Targets ... 63
    4.1 Introduction and Previous Results ... 64
        4.1.1 Previous Experimental Investigations ... 64
        4.1.2 Previous Theoretical and Numerical Investigations ... 67
    4.2 Experimental Campaign at CEA-Saclay ... 67
        4.2.1 Experimental Setup ... 68
        4.2.2 Experimental Results ... 73
    4.3 Numerical Simulations ... 79
        4.3.1 2D Simulation Campaign ... 80
        4.3.2 3D Simulation Campaign ... 82
    4.4 Theory of Surface Plasmon Acceleration ... 87
    4.5 Experimental Campaign at GIST ... 92
        4.5.1 Laser System ... 93
        4.5.2 Experimental Set-Up ... 93
        4.5.3 Preliminary Results ... 94
    4.6 Conclusions ... 95
    References ... 95

5 Foam Targets for Enhanced Ion Acceleration ... 99
    5.1 Introduction ... 100
        5.1.1 Requirements for a Laser-Based Ion Accelerator ... 100
        5.1.2 Previous Investigations with Foam Targets ... 102
    5.2 Experimental Activity ... 103
        5.2.1 Laser System ... 103
        5.2.2 Experimental Setup ... 104
        5.2.3 Targets ... 105
        5.2.4 Experimental Plan (First Campaign) ... 107
        5.2.5 Experimental Plan (Second Campaign) ... 107
    5.3 Experimental Results ... 108
        5.3.1 First Experimental Campaign: Enhanced Ion Acceleration ... 108
        5.3.2 Second Experimental Campaign: Effect of Pulse Length on Ion Acceleration with Foam Targets ... 111
    5.4 Numerical Simulations ... 112
        5.4.1 2D Simulations ... 112
        5.4.2 3D Simulations ... 121
        5.4.3 Modelling of Foam Target with Diffusion Limited Aggregation ... 125
    5.5 Conclusions ... 127
    References ... 128

6 Numerical Exploration of High Field Plasmonics in Different Scenarios ... 131
- 6.1 Rayleigh–Taylor Instability in Radiation Pressure Acceleration ... 132
  - 6.1.1 Radiation Pressure Acceleration ... 132
  - 6.1.2 Theoretical Model of Laser-Driven Rayleigh–Taylor Instability ... 136
  - 6.1.3 Numerical Simulations ... 143
  - 6.1.4 Conclusions ... 149
- 6.2 Plasmonic Effects in High Order Harmonic Generation from Grating Targets ... 150
  - 6.2.1 Introduction on HHG with Laser-Based Sources ... 150
  - 6.2.2 Grating Targets as a HHG Source ... 152
  - 6.2.3 Conclusions ... 157
- 6.3 Energy Concentration Schemes? ... 158
  - 6.3.1 Conclusions ... 160
- References ... 160

7 Conclusions and Perspectives ... 165
- Reference ... 166

Appendix A: Code Normalization ... 167

Appendix B: Particle-In-Cell algorithm ... 169

Curriculum Vitae ... 177

Abbreviations

BG/Q	BlueGene/Q microprocessor
CCD	Charge-coupled device
CWE	Coherent wake emission
DLA	Diffusion-limited aggregation
EM	Electromagnetic
EOS	Equation of state
HFP	High-field plasmonics
HHG	High-order harmonic generation
HPC	High-performance computing
MCP	Micro-channel plate
PF	Ponderomotive force
PIC	Particle-in-cell
PLD	Pulsed laser deposition
ROM	Relativistic oscillating mirror
RPA-HB	Radiation pressure acceleration—hole boring regime
RPA-LS	Radiation pressure acceleration—light sail regime
RTI	Rayleigh–Taylor Instability
SP	Surface plasmon
SW	Surface wave
TE	Transverse electric (mode)
TM	Transverse magnetic (mode)
TNSA	Target normal sheath acceleration
TPS	Thompson parabola spectrometer
XUV	eXtreme ultraviolet

Chapter 1
Introduction

Ultra-short laser-matter interaction at extreme intensities is a broad research field with several potential applications, including electron acceleration, ion acceleration, ultra-intense x-ray and γ sources, pulsed neutron sources and laboratory astrophysics (see Chap. 2). Laser-driven sources of charged particles or photons are characterized by some distinctive features with respect to conventional sources. Indeed, due to the shortness of the laser pulse, they can surpass the peak intensity of conventional sources. Moreover, the compactness of laser-based sources is another attractive feature: an "almost table-top" laser system can accelerate electrons up to a few GeVs [1] and produce γ rays with an intensity orders of magnitude higher than any conventional source (at least in some energy ranges [2]). Laser-based pulsed neutron sources have recently reached very high intensities [3–5] and their use is envisaged for active material interrogation or possibly even to study nucleosynthesis in conditions similar to those found in a supernova. Finally, laboratory astrophysics is gaining popularity because ultra-intense laser matter interaction allows to reach in the lab extreme pressures and temperatures [6] (similar to those found in the core of planets) or to study plasma instabilities relevant for astrophysics [7].

State-of-the art laser facilities can deliver laser pulses with a duration of tens of fs and an energy of tens of J, focused down to a few wavelengths spot sizes. This means that powers in excess of 1 PW and focused intensities $\lesssim 10^{22}$ W/cm^2 can be obtained. With laser intensities over 10^{15}–10^{16} W/cm^2 the electric field of the laser is high enough to suppress the Coulomb barrier in any atom (thus, in ultra-high intensity laser-matter interaction any target becomes rapidly a plasma). With laser intensities beyond 10^{18} W/cm^2 strongly relativistic effects start to play a role, since electrons acquire relativistic energies in a single field oscillation.

The development of laser technology in the past decades has driven the exploration of interaction regimes at increasingly high intensities and the up-coming 10 PW laser facilities may open further possibilities, ranging from the observation of Radiation Reaction effects[1] to ion acceleration up to 100s MeV and electron acceleration beyond 10 GeV.

The framework of the present manuscript is ultra-high intensity laser interaction with solid targets. In particular, the research activity presented here aims at exploring the role of plasmonic effects in several physical scenarios. A plasmon is a collective oscillation of the electrons in a plasma or a metal (the conduction electrons in a metal behave like a plasma), a surface plasmon is confined at the interface between the plasma or the metal and a dielectric, while a bulk plasmon can propagate in the volume. As will be detailed hereunder, plasmonic effects can arise spontaneously in some physical scenarios (possibly with detrimental effects) or the efficient coupling between the incoming laser and plasmons can be actively sought with suitable target configurations.

Three main research activities are presented in this manuscript: electron acceleration with relativistic surface plasmons induced with laser-grating interaction, enhanced ion acceleration with foam-attached targets and laser-driven Rayleigh–Taylor instability in radiation pressure ion acceleration. A short summary of the main findings and their relevance is provided hereunder.

As far as the first topic is of concern, the main result which we have obtained is the first observation of electron acceleration by relativistic surface plasmons excited with ultra-high intensity laser pulses interacting with grating targets [8, 9]. The study of surface plasmons in metallic and dielectric structures (plasmonics) at much lower field intensities is a vibrant research field, with several interesting research lines and applications [10]. However, the study of surface plasmons in a regime in which electromagnetic fields with relativistic intensities are involved is essentially an unexplored ground, without a complete theory. Actually, the mere existence of surface plasmons in these conditions, where strongly non-linear, relativistic effects are expected to take place, was by no means guaranteed. In "conventional" plasmonics, surface plasmons are typically excited with low-intensity laser pulses. Since the coupling between a surface plasmon and an electromagnetic wave is impossible with a flat interface between a metal and a dielectric, several coupling schemes have been developed; most of them require the use of dielectrics. These coupling schemes cannot be ported in the relativistic regime, since any material becomes a plasma (and hence a very good conductor) in a single laser cycle. There is however at least one coupling scheme which relies solely on a conductor and the vacuum. Indeed, irradiation of a periodically modulated conductor at definite resonance angles allows for the coupling between the incoming electromagnetic pulse and a surface plasmon. The main idea behind the experimental campaign performed at CEA-Saclay (Paris) in 2014 was exactly to test this scheme: grating and flat targets were irradiated with ultra-intense laser pulses ($I > 10^{18}$ W/cm^2) at various angles of incidence. A surface plasmon, which is expected to propagate almost at the speed of light if the plasma is dense enough, is characterized by strong longitudinal electric fields. Thus electrons

---

[1]An accelerated particle irradiates electromagnetic energy. The back-reaction force exerted on the particle due to this electromagnetic emission is called Radiation Reaction force.

extracted from the target might be trapped and accelerated by this longitudinal fields. Indeed, in the experimental campaign, collimated electron bunches close to the target tangent were observed when grating targets were irradiated close to the resonance angle expected for surface plasmon excitation. Their energy spectrum exhibited a peak at ∼5 MeV, with a tail extending up to ∼20–25 MeVs. The experimental results are supported by 3D particle-in-cell simulations and by a theoretical model. Besides the interest of this scheme as a pulsed, ultra-short, electron source,[2] these results could open the way for the extension of surface plasmon theory and possibly other plasmonic schemes into the unexplored ultra-high intensity regime. The encouraging preliminary results obtained recently in another laser facility with PW-class laser suggest that plasmonic effects can be observed and studied also at higher laser intensities.

Concerning instead the activity on foam attached-targets for enhanced ion acceleration, two experimental campaigns were performed at GIST laser facility, in Republic of Korea, where a PW-class system is available (a numerical investigation campaign was also carried out to support the experimental results). In previous experimental and numerical investigations [14, 15] a near-critical foam layer attached to a thin foil allowed for enhanced ion acceleration, if compared to simple flat foil (the enhancement concerns both the cut-off energy and the total number of accelerated particles). The near-critical foam layer is transparent for the pulse, which is able to penetrate for several wavelengths in the foam. This is the ideal scenario for an efficient energy transfer between the laser pulse and bulk plasmons. A bulk plasmon is a collective oscillation of the electrons in a plasma and it is characterized by a frequency ω_p. ω_p is a function of the electron density n_e ($\omega_p = \sqrt{4\pi n_e e_0^2/m_e}$, where m_e is the electron mass and e_0 is the elementary charge). An efficient energy transfer between the laser pulse and bulk plasmons can take place if the laser frequency $\omega \approx \omega_p$ (which is a resonance condition). The electron density which satisfies $\omega_p(n_e) = \omega$ is called "critical density" (n_c). If $\omega < \omega_p$ the laser field is evanescent in the plasma and thus the pulse cannot propagate. Instead, if $\omega > \omega_p$ the laser can propagate in the plasma. For solid density plasmas, $n_e > 100 n_c$, thus $\omega_p >> \omega$, preventing efficient energy transfer between the pulse and the bulk plasmons. The idea behind the use of foam-attached targets is that, if the foam has a density close to critical one, the laser is able to propagate through this near-critical layer, leading to an efficient absorption (i.e. more energy is transferred to the target electrons with respect to simple flat targets). Ion acceleration in this scenario is driven by the expansion of the heated electrons (thus if more energy is transferred to the the electrons, an enhanced acceleration process can be expected). Indeed, the expected enhancement was observed in the experimental campaign, together with a strongly reduced dependence on the polarization of the pulse (which strongly affects the interaction with

---

[2]Electron sources with these characteristics are not easily attainable with other techniques. Electron bunches in the few MeVs energy range could be interesting for imaging of ultra-fast processes with electron diffraction (i.e. Ultra-fast Electron Diffraction) [11–13] or photo-neutron generation (recently laser-based photoneutron sources were proven to reach very high peak flux intensities [3, 5]).

simple foils). With the aid of numerical simulations it was found that the structure of the foam plays a key role in reducing the effect of pulse polarization (the foam consists in nano-clusters at solid density arranged in complicated porous structures with a scalelength of the order of the laser wavelength). The study of foam-attached targets is interesting both to pursue the developments of advanced ion acceleration strategies and to investigate laser-plasma interaction in near-critical, nano-structured plasmas.

Finally, radiation pressure acceleration is a promising scheme for ion acceleration with ultra-intense laser pulses. In this regime, a very intense intense ($I \gtrsim 10^{22}$ W/cm^2), circularly polarized, laser pulse interacts with a thin foil (tens of nm). The laser acts directly on the electrons, pushing them forward. The intense electrostatic field generated by the displaced electrons is able to drag forward the ions of the target, so that the foil is accelerated as a whole. This scheme holds promise to allow laser-driven ion acceleration up to 100 s MeV energies. The intensity of present-day laser systems is not high enough to access a regime of "pure" radiation pressure acceleration, though some experimental evidence of this acceleration mechanism is already available. In these early experimental investigations, a net-like structure [16] in the spatial distribution of the particles is observed, suggesting the development of a Rayleigh–Taylor-like instability. Similar phenomena were observed also in numerical investigations [17]. In [18] we have studied in detail the development of this instability with numerical simulations and providing a theoretical model of the process. Plasmonic effects play a significant role in laser-driven Rayleigh–Taylor (RT) instability, since a resonant field enhancement takes place when the rippling of the target surface has a periodicity equal to the laser wavelength. In this case, indeed, the laser field (and thus the radiation pressure) is enhanced in the valleys of the rippling. Consequently, laser-driven RT favours the development of perturbations of the target surface with a scale-length close to the laser-wavelength. The laser ultimately breaks through the valleys of the rippling, leading to the observed net-like structures. The study of this process is important, since the instability may lead to an early onset of transparency, preventing the acceleration process.

This thesis is organized as follows:

- Chapter 2 provides a brief overview of the theoretical background of the present work (laser-matter interaction at extreme intensities, "traditional" plasmonics …). Also some technological aspects of laser development are discussed.
- In Chapter 3 the numerical tools developed for the research activity presented in this manuscript are discussed in detail. After an introduction on Particle-In-Cell simulations, the open-source PIC code *piccante* code and the spectral PIC code *piccolino* are described in detail and some applications are presented. Simulations performed with *piccante* play a central role in the research activities presented here.
- Chapter 4 presents the experimental and numerical activity on electron acceleration mediated by surface plasmons in laser-grating interaction at relativistic intensities. Also a theoretical model for the acceleration process is discussed.
- Chapter 5 discuss the experimental and numerical work on foam-attached targets.

- Chapter 6 is mainly devoted to the study the role of plasmonic effects in "light-sail" radiation pressure acceleration. In particular, the development of a laser-driven Rayleigh–Taylor instability is analysed in detail. In addition, the role of plasmonic enhancement of high-order harmonic generation with grating targets irradiated at resonance is discussed and some comments are provided on the possible extension of other plasmonic schemes in the ultra-high intensity regime.

I've been deeply involved in the research activities presented in this manuscript. I am among the two main developers of the open-source, massively parallel PIC code *piccante*, which proved to be essential for the computational needs of my group. I have also single-handedly developed a spectral PIC code (*piccolino*), which was especially useful for a secondary research project, described in Chap. 3. As far as the activity on surface plasmon acceleration is of concern, I contributed to the set-up of the experiment in CEA-Saclay (Paris), I took part to the whole experimental campaign (data collection and analysis), I performed all the numerical simulations and I contributed to the interpretation of the results. For the research activity on foam-attached targets, I took part to two experimental campaigns at GIST (Republic of Korea) and I was deeply involved in the numerical simulation campaign. Finally, I performed several numerical simulations for the study of laser-driven Rayleigh–Taylor instability and all the numerical simulations for the study of plasmonic enhancement of high-order harmonic generation.

References

1. W.P. Leemans, A.J. Gonsalves, H.-S. Mao, K. Nakamura, C. Benedetti, C.B. Schroeder, Cs. Tóth, J. Daniels, D.E. Mittelberger, S. S. Bulanov, J.-L. Vay, C.G.R. Geddes, E. Esarey, Multi-GeV electron beams from capillary-discharge-guided subpetawatt laser pulses in the self-trapping regime. Phys. Rev. Lett. **113**, 245002 (2014)
2. G. Sarri, K. Poder, J.M. Cole, W. Schumaker, A. Di Piazza, B. Reville, T. Dzelzainis, D. Doria, L.A. Gizzi, G. Grittani, S. Kar, C.H. Keitel, N. Krushelnick, S. Kuschel, S.P.D. Mangles, Z. Najmudin, N. Shukla, L.O. Silva, D. Symes, A.G.R. Thomas, M. Vargas, J. Vieira, M. Zepf, Generation of neutral and high-density electron–positron pair plasmas in the laboratory. *Nat. Commun.* **6**, 6747 (2015)
3. Y. Arikawa, M. Utsugi, A. Morace, T. Nagai, Y. Abe, S. Kojima, S. Sakata, H. Inoue, S. Fujioka, Z. Zhang, H. Chen, J. Park, J. Williams, T. Morita, Y. Sakawa, Y. Nakata, J. Kawanaka, T. Jitsuno, N. Sarukura, N. Miyanaga, H. Azechi, High-intensity neutron generation via laser-driven photonuclear reaction. Plasma Fusion Res. **10**, 2404003 (2015)
4. M. Roth, D. Jung, K. Falk, N. Guler, O. Deppert, M. Devlin, A. Favalli, J. Fernandez, D. Gautier, M. Geissel, R. Haight, C.E. Hamilton, B.M. Hegelich, R.P. Johnson, F. Merrill, G. Schaumann, K. Schoenberg, M. Schollmeier, T. Shimada, T. Taddeucci, J.L. Tybo, F. Wagner, S.A. Wender, C.H. Wilde, G.A. Wurden, Bright laser-driven neutron source based on the relativistic transparency of solids. Phys. Rev. Lett. **110**, 044802 (2013)
5. I. Pomerantz, E. McCary, A.R. Meadows, A. Arefiev, A.C. Bernstein, C. Chester, J. Cortez, M.E. Donovan, G. Dyer, E.W. Gaul, D. Hamilton, D. Kuk, A.C. Lestrade, C. Wang, T. Ditmire, B.M. Hegelich, Ultrashort pulsed neutron source. Phys. Rev. Lett. **113**, 184801 (2014)
6. M.A. Purvis, V.N. Shlyaptsev, R. Hollinger, C. Bargsten, A. Pukhov, A. Prieto, Y. Wang, B.M. Luther, L. Yin, S. Wang, J.J. Rocca, Relativistic plasma nanophotonics for ultrahigh energy density physics. Nat. Photonics **7**, 796–800 (2013)

7. É. Falize, A. Ravasio, B. Loupias, A. Dizière, C.D. Gregory, C. Michaut, C. Busschaert, C. Cavet, M. Koenig, High-energy density laboratory astrophysics studies of accretion shocks in magnetic cataclysmic variables. High Energy Density Phys. **8**(1), 1–4 (2012); cited By 7
8. L. Fedeli, A. Sgattoni, G. Cantono, I. Prencipe, M. Passoni, O. Klimo, J. Proska, A. Macchi, T. Ceccotti, Enhanced electron acceleration via ultra-intense laser interaction with structured targets. Proc. SPIE **9514**, 95140R–95140R-8 (2015)
9. L. Fedeli, A. Sgattoni, G. Cantono, D. Garzella, F. Réau, I. Prencipe, M. Passoni, M. Raynaud, M. Květoň, J. Proska, A. Macchi, T. Ceccotti, Electron acceleration by relativistic surface plasmons in laser-grating interaction. Phys. Rev. Lett. **116**, 015001 (2016)
10. Surface plasmon resurrection (editorial). Nat. Photonics **6**, 707 (2012)
11. J.B. Hastings, F.M. Rudakov, D.H. Dowell, J.F. Schmerge, J.D. Cardoza, J.M. Castro, S.M. Gierman, H. Loos, P.M. Weber, Ultrafast time-resolved electron diffraction with megavolt electron beams. Appl. Phys. Lett. **89**(18), 184109 (2006)
12. S. Tokita, S. Inoue, S. Masuno, M. Hashida, S. Sakabe, Single-shot ultrafast electron diffraction with a laser-accelerated sub-MeV electron pulse. Appl. Phys. Lett. **95**(11), 111911 (2009)
13. G. Sciaini, R.J.D. Miller, Femtosecond electron diffraction: heralding the era of atomically resolved dynamics. Rep. Prog. Phys. **74**(9), 096101 (2011)
14. M. Passoni, A. Zani, A. Sgattoni, D. Dellasega, A. Macchi, I. Prencipe, V. Floquet, P. Martin, T.V. Liseykina, T. Ceccotti, Energetic ions at moderate laser intensities using foam-based multi-layered targets. Plasma Phys. Control. Fusion **56**(4), 045001 (2014)
15. A. Sgattoni, P. Londrillo, A. Macchi, M. Passoni, Laser ion acceleration using a solid target coupled with a low-density layer. Phys. Rev. E **85**, 036405 (2012)
16. C.A.J. Palmer, J. Schreiber, S.R. Nagel, N.P. Dover, C. Bellei, F.N. Beg, S. Bott, R.J. Clarke, A.E. Dangor, S.M. Hassan, P. Hilz, D. Jung, S. Kneip, S.P.D. Mangles, K.L. Lancaster, A. Rehman, A.P.L. Robinson, C. Spindloe, J. Szerypo, M. Tatarakis, M. Yeung, M. Zepf, Z. Najmudin, Rayleigh–Taylor instability of an ultrathin foil accelerated by the radiation pressure of an intense laser. Phys. Rev. Lett. **108**, 225002 (2012)
17. F. Pegoraro, S.V. Bulanov, Photon bubbles and ion acceleration in a plasma dominated by the radiation pressure of an electromagnetic pulse. Phys. Rev. Lett. **99**, 065002 (2007)
18. A. Sgattoni, S. Sinigardi, L. Fedeli, F. Pegoraro, A. Macchi, Laser-driven Rayleigh–Taylor instability: plasmonic effects and three-dimensional structures. Phys. Rev. E **91**, 013106 (2015)

Chapter 2
Introduction on High Intensity Laser-Plasma Interaction and High Field Plasmonics

This dissertation deals manly with the attempt to extend the study of plasmonic effects in the ultra-high intensity (beyond 10^{18} W/cm^2) laser-matter interaction regime. Plasmonics, which is the study of surface plasmons, is a mature research field. However, surface plasmons are generally excited with low-intensity laser pulses. The study of plasmonic effects when ultra-high intensity lasers are involved is an almost completely unexplored ground. In this regime, which will be referred as *High Field Plasmonics* in the following, relativistic, strongly non-linear effects are expected to take place.

This chapter is intended to provide a concise overview on laser-matter interaction in this intensity regime and to introduce the topic of High Field Plasmonics.

Section 2.1 provides an overview of high intensity laser technology. In Sect. 2.2 we give a brief introduction on the theory of laser-matter interaction in the high field regime, characterized by relativistic and strongly non-linear effects. Finally, Sect. 2.3 provides a synthetic theoretical introduction on what is meant for High Field Plasmonics, highlighting in particular possible future developments and foreseen difficulties.

2.1 Evolution of High Intensity Laser Technology

Since the realization of the first laser[1] in 1960 [3], laser technology underwent a tremendous development. An impressive and yet unbroken series of innovations has led to the realization of a wide variety of laser systems with very different features,

---

[1]The first Laser was preceded by the first Maser in 1953 [1], operating in the microwave region of the electromagnetic spectrum. Basic laser physics is beyond the scope of this dissertation and the interested reader is referred to [2].

enabling a broad range of applications [4], so that today the use of lasers is pervasive in both science and technology.

In this document, we take a very partial view on the history of lasers: the brief overview given in Sect. 2.1.1 will be limited to ultra-high intensity laser systems.

2.1.1 Overview

Figure 2.1 shows synthetically the development of high intensity laser systems over time. Starting from 1960, intensity increased at a fast pace for approximately a decade. This evolution was marked by two main innovations: Q-switching [7] and mode-locking [8]. These schemes allowed to increase the peak intensity of laser irradiation by orders of magnitude. Q-switching strategy is based on the sudden variation of the quality factor of the resonator cavity containing the laser active medium (this allows for more electrons to reach the excited state). The mode-locking strategy instead relies on the phase coherence of the cavity modes in the resonator. Both schemes result in a pulsed rather than a continuous emission.

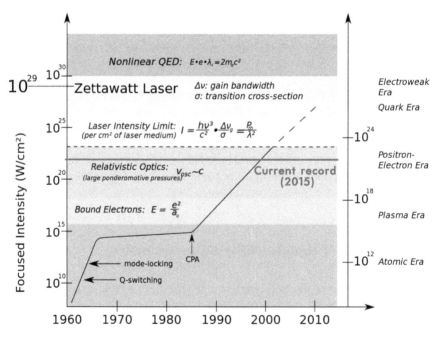

Fig. 2.1 In this graph, the maximum intensity reached by laser systems is shown as a function of time. The main innovations leading to higher intensities are highlighted. The *red horizontal curve* represents the current record (see [5]). The terminal part of the trend line (the *black curve*) is based on estimations formulated around 2000, which have proven to be over-optimistic. The picture was adapted from [6] (Color figure online)

2.1 Evolution of High Intensity Laser Technology

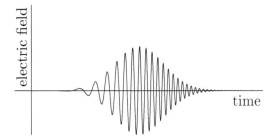

Fig. 2.2 This graph illustrates the electric field as a function of time for a chirped pulse. The instantaneous frequency of the pulse is evidently a function of time. In this case the instantaneous frequency grows linearly with time and the pulse is called *upchirped*

In order to reach high intensities, the laser emission of the resonator cavity can be fed into a sequence of amplifying stages. Indeed, if coherent radiation is made to pass through an excited active medium, it stimulates further coherent radiation emission. Thus a laser pulse can be coherently amplified extracting energy from the active medium.

Of course the possibility to reach extremely high intensities is strongly linked to the pulse duration. Presently the highest peak power of a laser system is of the order of 1 PW, which can be sustained at most for few ps or tens of ps (most PW class laser system have a pulse duration of only tens of fs).[2] The requirement on pulse duration is a strong limit for the active medium since the well known inequality $\Delta \tau \Delta \omega \geq 1/2$ forces the active medium to operate in a large frequency range $\Delta \omega$ (i.e. it should have a large *gain bandwidth*) in order to have a short $\Delta \tau$. Three main technologies are exploited for high intensity laser systems: Titanium:Sapphire [10] lasers (the active medium is a Sapphire, Al_2O_3, crystal doped with Titanium ions), CO_2 lasers [11–13] (with a gaseous active medium containing CO_2) and Nd:YAG lasers [10] (based on neodimium doped yttrium-aluminium garnet crystal). The shortest pulse durations which can be obtained with Nd:YAG lasers and CO_2 lasers are, respectively, in the ps range and in the 100 fs range, whereas Ti:Sapphire lasers can provide pulses with durations down to ∼10 fs. The record peak powers for Nd:YAG and Ti:Sapphire systems are in the PW range, while CO_2 lasers are approaching the 100 TW milestone. Here we will be mainly concerned with Ti:Sapphire laser systems, since for High Field Plasmonics very short pulse durations are needed.

Despite the rapid growth in the sixties, as shown in Fig. 2.1 the maximum focused intensity remained flat for decades, until mid-eighties, when *Chirped Pulse Amplification* (see [14]) opened the way for a new fast-pace increase of peak focused intensities of laser systems.

For long time further improvements of laser peak intensities were prevented by the material limits of the laser chain. Indeed, increasing the pulse intensity would have damaged the optical components. Moreover, the amplification process itself is inefficient for very short, intense pulses. Chirped Pulse Amplification (CPA) allowed to overcome these difficulties.

---

[2] For a comparison, the average electrical power produced on Earth is ∼2–3 TW [9]. Thus a PW class laser system is hundreds of times more powerful than the total electric output of the Earth.

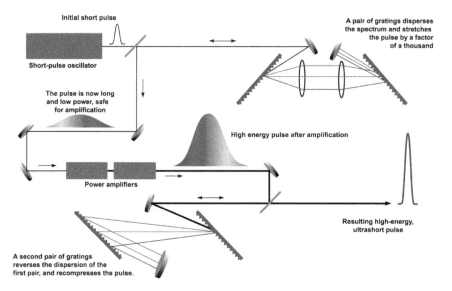

Fig. 2.3 A typical implementation of CPA is show. In this case the pulse stretching is obtained with a couple of gratings. After the first stage, the long chirped pulse is amplified through a series of power amplifiers. Finally the pulse is compressed again with a couple of gratings. These last gratings are typically very large, in order to avoid damage. This last stage of the optical chain is called *compressor*. The figure is reproduced from [15]

CPA consists essentially in increasing the temporal length of laser pulses introducing a chirp (see Fig. 2.2): spectral components of the pulse are dispersed in time according to their frequency (this *stretching* can be achieved with a couple of gratings or with a long optical fibre). A laser pulse whose spectral components are dispersed in time according to their frequency is called *chirped*.[3] A stretched pulse is easier to amplify efficiently and with lower damage risks for the optical components. Dispersion of spectral components can be reversed after the amplification stages with a couple of gratings in the *compressor* stage. In Fig. 2.3 a scheme of a typical CPA system is illustrated.

Currently, the record focused peak intensity for modern CPA-based laser systems is 2×10^{22} W/cm^2, reached by a 300 TW Ti:Sapphire laser (see [5]).

2.1.2 A Typical High Intensity Ti:Sapphire Laser System

A typical High Intensity Ti:Sapphire laser system is roughly described by Fig. 2.3. An oscillator is responsible for producing a high-quality short pulse. These devices operate usually at a very high frequency ($\sim$ MHz), far beyond the capabilities of the

---

[3] An audio signal with similar properties recalls the chirping of birds.

2.1 Evolution of High Intensity Laser Technology

other stages of the chain,[4] thus a single pulse is selected for further amplification. The pulse is then stretched and amplified in a sequence of stages. Typically, the active medium of these stages is efficiently *pumped* (i.e. excited) with other lasers, in order to prevent excessive heating. The amplified chirped pulse is finally compressed with a couple of gratings and sent to the interaction chamber, where the laser is focused on a target with a parabolic mirror.

Since several amplification stages are usually involved, care should be taken in order to avoid *gain narrowing*, which could increase pulse duration.

The laser pulse sent to the experimental room usually needs further optimizations. In particular, an inherent issue of high-power laser systems is the emission of intense prepulses both on the ns and the ps time scale before the main pulse. These pre-pulse, though sometimes useful, are typically detrimental for the experimental activity. Especially for High Field Plasmonics, for which the laser is made to interact with a solid structured target, a careful control of pre-pulses is needed. Control of pre-pulses is usually performed with plasma mirrors [16, 17]: the laser is focused on a transparent medium and the prepulses are either transmitted or absorbed. The plasma mirror is designed in such a way that a plasma is formed just before the main pulse, which is reflected with moderate losses. Most modern systems use a double plasma mirror. At the price of loosing ∼50 % of the initial pulse energy, with a double plasma mirror a contrast of 10^{10} can usually be achieved.

Finally, adaptive optics techniques are usually needed for focal spot optimization, in order to reach the highest possible focused intensities.

2.1.3 Towards 10 PW Laser Systems

Several projects to design and build ultra-intense laser systems able to overcome the PW limit are currently ongoing. A recent review on the current status and foreseen developments for ultra-high intensity laser technology can be found in [18]. PULSER laser at GIST (Gwangju, Republic of Korea) should soon operate in the multi-petawatt regime (4 PW are expected), since the upgrade from 1 PW is already in progress [19].

Apollon laser is being built in France with a target power of ∼10 PW [20]. Though the construction works have been delayed, also RAL-CLF facility (UK) has prepared plans for a 10 PW upgrade of Vulcan laser [21] and is currently involved in the validation of the concept. Both Apollon project and Vulcan upgrade project are based on on the Optical Parametric Chirped Pulse Amplification technique (OPCPA, see [22–24]).

Finally, in 2014, the large European project ELI (Extreme Light Infrastructure) has awarded a contract for the development and delivery of a 10 PW system to an industrial and academic consortium and also the Shanghai Institute of Optics and fine Mechanics (SIOM) is currently working on a 10 PW laser system [25].

---

[4]Ti:Sapphire laser systems operate in the 1–10 Hz frequency range (i.e. one shot every 0.1–1 s), mainly due to issues related to the amplifying stages.

If compared to the rapid development of high-intensity laser technology in the past decades, achieving the 10 PW milestone is requiring an amount of time significantly greater than expected. This might suggest that we are approaching the fundamental limits of current technology. However, some radically different schemes are currently being developed, tough being still a concept at the moment. An example is the multi-fibre laser, whose development is currently pursued by the International Coherent Amplification Consortium. This laser system should consist in a large bundle of fibre-based lasers which are coherently combined to achieve extremely high intensities with a reasonable efficiency.[5] If the technical issues will be tackled, the proponents claim that this scheme could provide an efficient, high repetition rate 100 PW laser[6] concept (see [28, 29]).

2.2 Relativistic Laser Plasma Interaction

This section is intended to provide an introduction on Laser-Plasma interaction at ultra-high intensities. A natural starting point is single particle motion in a relativistic laser field. The study of ionization processes, which would be very important for lower laser intensities ($I < 10^{16}$ W/cm^2), is completely ignored here: at relativistic intensities, at least the outer electron shells of any atom are ionized within one laser cycle[7] [30]. Section 2.2.1 deals with single particle motion in intense laser fields. Electromagnetic (EM) waves propagation in a plasma is discussed in Sect. 2.2.2. Kinetic theory in the relativistic regime is briefly introduced in Sect. 2.2.3, whereas the importance of intense irradiation due to charged particle motion and QED effects at extremely high intensities are discussed in Sect. 2.2.6. Finally, possible applications and foreseen future developments of laser-plasma interaction are covered in Sect. 2.2.7.

The reader interested in Laser-Plasma interaction is referred to [30–32] which cover the topic extensively. For the Relativistic Kinetic theory a comprehensive reference is provided by [33].

2.2.1 Single Particle Motion

We start with a simple derivation of the non-relativistic *quiver motion* of an electron in an oscillating electric field. The equation of motion is simply:

---

[5]Fibre-based lasers can operate in the kW regime with a wall-plug efficiency $\sim 30\%$.

[6]Possible applications of a laser built according to this concept include space debris control [26] and production of Tc 99 m [27], a radioactive isotope of medical interest.

[7]The critical intensity for ionization of any material with single photon processes is $\sim 3.5 \times 10^{16}$ W/cm^2.

2.2 Relativistic Laser Plasma Interaction

$$\frac{d^2x}{dt^2} = -\frac{e_0}{m} E_0 \cos(\omega t) \qquad (2.1)$$

We can integrate once Eq. 2.1, which leads to the following expression for the velocity as a function of time:

$$v = \frac{e_0 E_0}{m\omega} \sin(\omega t) \qquad (2.2)$$

we define a_0 as:

$$a_0 = \frac{v}{c} = \frac{e_0 E_0}{m\omega c} \qquad (2.3)$$

The adimensional a_0 parameter is crucial in Laser-Plasma interaction, since its value characterises the interaction properties: $a_0 \ll 1$ means no relativistic effects, $a_0 \lesssim 1$ is a regime with weakly relativistic effects, $a_0 \geq 1$ defines the fully relativistic regime.

Equation 2.2 is oversimplified, since the effect of the magnetic field component is completely neglected. If $v \ll c$ this approximation is justified: the strength of the force exerted by the magnetic field component is indeed suppressed by a factor v/c. However, if $v \approx c$, as in relativistic laser-plasma interaction, the magnetic and the electric field components become equally important. An exact solution of electron motion in an electromagnetic plane wave field can be found and the derivation is reported hereunder (closely following [30] and [31]).

The equations for particle momentum $\mathbf{p}$ and energy $E = m_e \gamma c^2$ read as follows:

$$\frac{d\mathbf{p}}{dt} = -e_0 \left(\mathbf{E} + \frac{\mathbf{v}}{c} \times \mathbf{B} \right) \qquad (2.4)$$

$$\frac{d}{dt}\left(m_e \gamma c^2\right) = -e_0 \mathbf{v} \cdot \mathbf{E} \qquad (2.5)$$

with $\mathbf{p} = \gamma m \mathbf{v}$ and $\gamma = \sqrt{1 + p^2/m^2 c^2}$.

We consider an elliptically polarized plane-wave, described by the vector potential $\mathbf{A}$, travelling along $\hat{\mathbf{x}}$ ($\mathbf{k} = k\hat{\mathbf{x}}$):

$$\mathbf{A}(\omega, \mathbf{k}) = (0, \delta a_0 \cos \phi, \sqrt{1 - \delta^2} a_0 \sin \phi) \qquad (2.6)$$

where $\phi = \omega t - kx$ is the phase. The δ parameter controls the pulse polarization ($\delta = 0, \pm 1$ means P-polarization, while $\delta = \pm 1/\sqrt{2}$ means C-polarization).

Since $\mathbf{E} = -\frac{1}{c}\partial_t \mathbf{A}$ and $\mathbf{B} = \nabla \times \mathbf{A}$ (equal to $\hat{\mathbf{x}} \times \partial_x \mathbf{A}$) we can re-write Eq. 2.4 as:

$$\frac{dp_x}{dt} = -e_0 \left(\partial_t \mathbf{A} + \frac{\mathbf{v}}{c} \times \nabla \times \mathbf{A} \right)\big|_x \qquad (2.7)$$

$$\frac{d\mathbf{p}_\perp}{dt} = -e_0 \left(\partial_t \mathbf{A} + \frac{\mathbf{v}}{c} \times \nabla \times \mathbf{A} \right)\big|_\perp \qquad (2.8)$$

where the momentum $\mathbf{p}$ is broken down into its longitudinal and transversal components. Using the fact that $\mathbf{A} = \mathbf{A}_\perp$ (i.e. $\mathbf{A}$ doesn't have a longitudinal component) and that $\mathbf{v} \times \nabla \times \mathbf{A} = -v_x \partial_x \mathbf{A}_\perp$, from Eq. 2.8 we get:

$$\frac{d}{dt}\left(\mathbf{p}_\perp - \frac{e}{c}\mathbf{A}\right) = 0 \qquad (2.9)$$

which states the conservation of canonical momentum (related to translational invariance in the transverse plane).

It can be shown that a second conservation law can be written, replacing $\mathbf{A}$ in Eq. 2.5:

$$\frac{d}{dt}(p_x - m_e \gamma c) = 0 \qquad (2.10)$$

so that the particle motion is completely determined by the two constants of motion $\mathbf{p}_\perp - \frac{e}{c}\mathbf{A}$ and $p_x - m_e \gamma c$.

Considering that the electron is initially at rest leads to the following solution, which depends parametrically on δ:

$$\begin{cases} \dfrac{\hat{x}}{a_0^2} = \dfrac{1}{4}\left[-\phi - \left(\delta^2 - \dfrac{1}{2}\right)\sin 2\phi\right] \\ \dfrac{\hat{y}}{a_0} = -\delta \sin(\phi) \\ \dfrac{\hat{z}}{a_0} = (1 - \delta^2)^{1/2} \cos(\phi) \end{cases} \qquad (2.11)$$

where the coordinates have been normalized with respect to $1/k$. Figure 2.4 shows the trajectory of an electron in intense, P-polarized and C-polarized laser pulses. We immediately notice that in both cases there is a systematic drift of the electron along $\hat{x}$. It is evident from the graph that for P-polarization the electron is confined on the xy plane, while for C-polarization it spirals around the $\hat{x}$ axis.

2.2.2 Propagation of EM Waves in a Plasma

After having derived the behaviour of a single electron in an intense oscillating EM field, the natural prosecution is the study of the propagation of EM waves in plasmas. For simplicity only the propagation of EM waves in cold, non-magnetized plasmas will be considered.[8]

The equation which describes the propagation of an EM wave reads as follows:

---

[8] The interested reader can found a thorough discussion of the rich physics of EM waves propagation in plasmas in [34].

2.2 Relativistic Laser Plasma Interaction

Fig. 2.4 Trajectory of a charged particle under the effect of a relativistic C-polarized (upper panel) or P-polarized laser pulse

$$\left(\nabla^2 - \frac{1}{c^2}\partial_t^2\right)\mathbf{E} - \nabla(\nabla \cdot \mathbf{E}) = \frac{4\pi}{c^2}\partial_t \mathbf{J} \qquad (2.12)$$

The derivation of the dispersion relation for EM waves in a cold plasma can be found in several standard textbooks [34, 35], thus only the result of the derivation will be reported here:

$$-k^2 c^2 + \epsilon(\omega)\omega^2 = 0 \qquad (2.13)$$

where

$$\epsilon(\omega) = 1 - \frac{\omega_p^2}{\omega^2} \qquad (2.14)$$

In the previous equation, ω_p is the *plasma frequency*, defined as follows:

$$\omega_p = \sqrt{\frac{4\pi e^2 n_e}{m_e}} \qquad (2.15)$$

Consequently, when $\omega > \omega_p$ an EM wave is free to propagate in the plasma, whereas if $\omega < \omega_p$ the EM wave is exponentially damped in the plasma. Given a fixed EM frequency ω the *critical density* n_c is defined as the plasma density which makes $\omega_p = \omega$. For a laser wavelength ~800 nanometers,[9] the critical density $n_c \sim 2 \cdot 10^{22}$ cm^{-3}.

The previous derivation is valid for a non-relativistic case. Relativistic laser-plasma interaction is significantly more challenging to study. Indeed, the strongly

---

[9]As for a typical Ti:Sapphire laser system.

non-linear effects lead to the complicated regime of non-linear optics.[10] However, at least for circularly polarized EM waves, an exact solution exists, provided that finite time and finite space effects are disregarded. In this last case the correct dispersion relation can be obtained with the replacement $m_e \rightarrow \gamma_e m_e$, where $\gamma_e = \sqrt{1 + a_0^2/2}$. This means that, in general, for a given frequency w, relativistic effects raise the critical density n_c to $\gamma_e n_c$. This phenomenon is called *relativistic transparency*.

2.2.3 Relativistic Kinetic Equations

In several physical scenarios, kinetic effects cannot be neglected. Here a synthetic derivation of relativistic kinetic equations is given (a comprehensive treatment can be found in [33]).

The relativistic one-particle distribution function $f(x, p)$ is the probability of finding a particle within a small $\Delta^4 x$ around four-position x and a small $\Delta^4 p$ around four-momentum p (following reference [33], a classical picture is adopted, thus all quantum effects are neglected).

Given a collection of identical particles with mass m, $f(x, p)$ is defined as a statistical average of $\sum_{i=1}^{N} \delta^3 (\mathbf{x} - \mathbf{x}_i(t)) \delta^3 (\mathbf{p} - \mathbf{p}_i(t))$, where $\mathbf{x}_i(t)$ and $\mathbf{p}_i(t)$ are, respectively, the three-position and the three-momentum of particle i at time t[11]:

$$f(x, p) = \left\langle \sum_{i=1}^{N} \delta^3 (\mathbf{x} - \mathbf{x}_i(t)) \delta^3 (\mathbf{p} - \mathbf{p}_i(t)) \right\rangle \quad (2.16)$$

Using $f(x, p)$, particle four-flow density can be defined as:

$$N^\mu = c \int \frac{d^3 p}{p^0} f(x, p) \quad (2.17)$$

We will now derive the collisionless kinetic equation for the evolution of $f(x, p)$ without any external force. We can define the scalar quantity $\Delta N(x, p)$ as follows:

---

[10] Phenomena like self-focusing, higher frequency generation ...may take place and concepts like the refraction index and the dispersion relation cannot be ported straightforwardly in this regime.

[11] Though not manifestly covariant, $f(x, p)$ can be proven to be a Lorentz scalar using the ancillary function $\mathcal{N}(x, p) = \frac{1}{p^0} \delta(p^0 - \sqrt{p^2 + m^2 c^2}) f(x, p)$. It is possible to show that $\mathcal{N}(x, p) = \frac{1}{mc} \int d\tau \langle \sum_{i=1}^{N} \delta^4 (x - x_i(t)) \delta^4 (p - p_i(t)) \rangle$, which is a Lorentz scalar (τ is the proper time). Moreover, it is trivial to show that $\theta(p^0) \delta(p^\mu p_\mu - m^2 c^2) = \frac{\delta(p^0 - \sqrt{p^2 + m^2 c^2})}{2 p^0}$. Finally, since using the previous result $\mathcal{N}(x, p) = \theta(p^0) \delta(p^\mu p_\mu - m^2 c^2) f(x, p)$, we can conclude that $f(x, p)$ is a Lorentz scalar.

2.2 Relativistic Laser Plasma Interaction

$$\Delta N(x, p) = \int_{\Delta^3 \sigma} \int_{\Delta^3 p} d^3\sigma_\mu \frac{d^3 p}{p_0} p^\mu f(x, p) \quad (2.18)$$

where $d^3\sigma_\mu$ (time-like four-vector) is an oriented three-surface element of a plane space-like surface. $\Delta N(x, p)$ can be interpreted as the number of world lines crossing the surface element $\Delta^3 \sigma$ with a four-momentum within a volume $\Delta^3 p$ centred on p^μ.

After some time, the same number of world lines will cross the surface element $\Delta^3 \hat{\sigma}$. Thus we get:

$$\int_{\Delta^3 \hat{\sigma}} \int_{\Delta^3 p} d^3\sigma_\mu \frac{d^3 p}{p_0} p^\mu f(x, p) - \int_{\Delta^3 \sigma} \int_{\Delta^3 p} d^3\sigma_\mu \frac{d^3 p}{p_0} p^\mu f(x, p) = 0 \quad (2.19)$$

The bundle of word lines crossing $\Delta^3 \sigma$ and $\Delta^3 \hat{\sigma}$ defines a word line tube, whose enclosed four-volume is $\Delta^4 x$. Since there are no collisions, world lines cannot cross the borders of the tube. Exploiting Gauss theorem we can write

$$\int_{\Delta^4 x} \int_{\Delta^3 p} d^3\sigma_\mu \frac{d^3 p}{p_0} p^\mu \partial_\mu f(x, p) = 0 \quad (2.20)$$

which immediately leads to

$$p^\mu \partial_\mu f(x, p) = 0 \quad (\partial_t + \mathbf{u} \cdot \nabla) f(x, p) \quad (2.21)$$

Equation 2.21 simply describe the free-streaming of the distribution function ($\mathbf{u}$ is the three-velocity $\mathbf{u} = \mathbf{p}/(\gamma m_0)$).

A very similar derivation can be carried out also if an external force F^μ is present. In this case, a particle following a world line between $\Delta^3 \sigma$ and $\Delta^3 \hat{\sigma}$ changes momentum as follows: $p^\mu \to p^\mu + F^\mu \Delta \tau$, where τ is the proper time.

Omitting the full derivation (details can be found in [33]), the final result for the relativistic kinetic equation is:

$$p^\mu \partial_\mu f(x, p) + m F^\mu \frac{\partial}{\partial p^\mu} f(x, p) = 0 \quad (2.22)$$

Equation 2.22 is valid only if F^μ doesn't alter the rest mass m_0 of the particles ($p^\mu F_\mu = 0$) and if $\frac{\partial F^\mu}{\partial p^\mu} = 0$. It is worth to mention that these conditions are satisfied for the electromagnetic force $F^\mu = -\frac{q}{mc} \mathcal{F}^{\mu\nu} p_\nu$ (where $\mathcal{F}^{\mu\nu}$ is the electromagnetic tensor).

Equation 2.22 can be rewritten as

$$(\partial_t + \mathbf{u} \cdot \nabla + \mathbf{F} \cdot \frac{\partial}{\partial p}) f(x, p) = 0 \quad (2.23)$$

2.2.4 Energy Absorption with Overdense Targets

In the previous subsection we have shown that an EM wave cannot penetrate in an overdense plasma. When fully ionized, solid density targets are strongly overdense ($n_e/n_c \gg 100$) for Ti:Sapphire fs lasers. This means that a laser pulse is usually reflected from these targets, unless they are so thin that the laser is able to break trough them.

Even if the laser is reflected back, a significant fraction of its energy may be transferred to the electrons of the target. Indeed "fast electrons" with energies of a few MeVs can usually be observed for ultra-high intensity laser-matter interaction. The typical order of magnitude of the hot electron energies as a function of the a_0 parameter can be estimated[12] as $E_{hot} = m_e c^2 \left(\sqrt{1 + a_0^2/2} - 1\right)$, so that laser-solid interaction with a $a_0 = 5$ (typical for 100 TW laser systems) the expected energy for the hot electron emission is ~ 1.4 MeV. Depending on the irradiation conditions, several heating processes exist and the most relevant for ultra-high intensity laser interaction with solid density plasmas are listed below (see [31, 36] for a more detailed discussion). Electron heating is strongly dependent on the pulse parameters (polarization, angle of incidence, intensity) and on the target properties (density of the plasma, steep interface or smooth density gradient).

The study of electron heating processes is important because the expansion of the heated electrons leads to very intense electrostatic fields, which in turn can accelerate ions (see Sect. 2.2.7). Since high energy ion sources are of interest for several applications, understanding and controlling these processes may be beneficial for the optimization of laser-based ion sources.

Resonance Absorption

A possible mechanism for electron heating in laser-plasma interaction is *resonance absorption*, which can take place in a region of the plasma where the plasma frequency ω_p is equal to the laser frequency ω. Indeed, in these conditions, the laser can efficiently couple with the normal modes (bulk plasmons) of the plasma, which propagate with a frequency ω_p. Of course, if a solid target is (even partially) ionized, $\omega_p \gg \omega$. However, intense laser pulses are frequently preceded by a less intense prepulse on the ps (or even ns) timescale. If the pre-pulse is sufficiently intense, it may lead to a pre-expansion of the target, forming a density gradient. The resonance absorption mechanism can then take place in the region of this plasma gradient in which the resonance condition is satisfied.

A simple electrostatic model of the process is described in [31] and is reported here.

The EM wave is described as an external oscillating field $\mathbf{E} = \mathfrak{Re}\left(\tilde{\mathbf{E}}_d e^{-i\omega t}\right)$. The EM wave propagates in a plasma with a background density which is a function of the position $n_0 = n_0(x)$. The following system of equations can be written for this physical scenario, combining Gauss's law for the electric field with fluid equations

---

[12] This estimated energy is called the "ponderomotive energy".

2.2 Relativistic Laser Plasma Interaction

for the plasma:

$$\begin{cases} \nabla \cdot \mathbf{E} &= -4\pi e(n_e - n_0) \\ \partial_t n_e &= -\nabla \cdot (n_e \mathbf{u}) \\ (\partial_t - \mathbf{u} \cdot \nabla)\mathbf{u} &= -\dfrac{e}{m_e}(\mathbf{E} + \mathbf{E}_d) \end{cases} \quad (2.24)$$

Linearising the previous equations in the limit $u_x \ll L/\omega$ (i.e. density is considered to be uniform over the electron oscillation amplitude) we get:

$$\delta n_e = \frac{1}{4\pi e} \frac{(\mathbf{E} + \mathbf{E}_d) \cdot \nabla n_0}{n_0(x) - n_c} \quad (2.25)$$

Equation 2.25 indicates that there is a resonance when the plasma density is equal to the critical density, unless $\mathbf{E} \cdot \nabla n_0 = 0$. This means that P-polarization and oblique incidence are required. Actually, the previous derivation should be modified for oblique incidence (the laser pulse is reflected back at a density lower than n_c, possibly preventing the coupling with bulk plasmons). A detailed analysis of this effect is however beyond the scope of this section.

Vacuum Heating

Vacuum heating is an electron absorption process which takes place in high intensity laser-plasma interaction when steep plasma gradients are involved. This mechanism was first proposed by Brunel [37]. Essentially the combined electric field of the incident pulse and the reflected pulse is responsible for dragging the electrons out of the target. After a half-oscillation, the electrons are injected in the target at high energy (the oscillation energy) and, since the EM field is evanescent in the overdense plasma, they are free to propagate further in the target. This mechanism should produce a high-energy electron bunch every laser cycle (the properties of these bunches can be studied as described in [38]).

A minimal 1D model of the process can be provided (see [31, 36, 39]), considering a step-like plasma with ion density $n_i = n_0 \Theta(x)$ and an EM field $E = E_e + E_d$, where E_e is the electrostatic component and E_d is the oscillating external driver field $E_d = E_0 \sin(\omega t)$ (only the $\hat{x}$ component of the fields is considered).

Adopting a fluid description of the plasma, the following equations can be written:

$$\begin{cases} \partial_x E_e &= 4\pi \rho = 4\pi e \left[n_0 \Theta(x) - n_e \right] \\ \partial_t n_e &= -\partial_x (n_e v_x) \\ \dfrac{dv_x}{dt} (\partial_t + v_x \partial_x) v_x &= -\dfrac{e}{m_e}(E_e + E_d) \end{cases} \quad (2.26)$$

As described in [31], switching to Lagrangian variables ($x = x_0 + \xi$, $\tau = t$ and $d\xi/dt = v_x$) leads to the following solution:

$$\xi = -x_0 + u_d (\sin(\omega t) - \sin(\omega t_0)) - \frac{u_d \cos \omega t_0}{1 - (\omega/\omega_p)^2}(t - t_0) + \frac{u_d \omega}{2}(t - t_0)^2 \sin(\omega t_0) \quad (x_0 + \xi < 0)$$

$$(2.27)$$

where $u_d = e\tilde{E}_e/(m_e/\omega)$. The last term of Eq. 2.27 shows that the electrons experience a secular acceleration (they enter the plasma with a velocity $\approx u_d$). In the relativistic regime, a reasonable approximation of the kinetic energy of the re-entrant electrons is given by the ponderomotive energy. The third term of Eq. 2.27 instead shows that there is a resonance when $\omega \sim \omega_p$.

Vacuum heating relies crucially on longitudinal field components, which are absent for S-polarized laser pulses or for normal incidence.

It can be shown (see [30] for the detailed derivation) that a good estimation of the laser absorption efficiency with the vacuum heating process (in the limit $\omega \ll \omega_p$, far from the resonance) can be obtained solving the following implicit equation:

$$\eta = \frac{1+\sqrt{1-\eta}}{\pi a_0}\left[\left(1+a_0^2\sin^2\theta(1+\sqrt{1-\eta})^2\right)^{1/2}-1\right]\frac{\sin\theta}{\cos\theta} \quad (2.28)$$

where η is the absorption efficiency and θ is the angle of incidence of the pulse. Equation 2.28 is obtained assuming that the energy of the re-entrant electrons is the ponderomotive energy and taking into account self-consistently the reduction of the amplitude of the accelerating field due to the absorption process. A few solutions of Eq. 2.28 are represented in Fig. 2.5. For small a_0 the absorption efficiency peaks at grazing incidence, while for large a_0 the model predicts η to reach 1 at smaller angles of incidence.

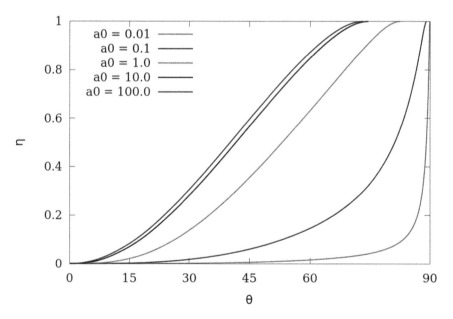

Fig. 2.5 The solutions of Eq. 2.28 are shown for several a_0, ranging from $a_0 = 0.01$ (non-relativistic regime) to $a_0 = 100$ strongly relativistic regime

J × B Heating

J × B heating mechanism [40] is similar to vacuum heating, since it requires a very steep plasma gradient. While in vacuum heating the electrons are extracted from the target and accelerated in the vacuum by the electric field of the pulse, **J × B** relies on the **v × B** term of the Lorentz force. At normal incidence, **J × B** can be considered as a form of vacuum heating where E_d is replaced by $\mathbf{v} \times \mathbf{B}/c$. The main difference between the two processes is that E_d varies with a frequency ω, while the frequency of $\mathbf{v} \times \mathbf{B}/c$ is 2ω.

It can be shown that the longitudinal force exerted via **J × B** mechanism is proportional to $(1 - \cos(2\omega t))$. The constant term is simply the ponderomotive force, while the oscillating term is responsible for accelerating electron bunches into the target with a frequency equal to 2ω. It can be shown that **J × B** heating has a resonance for $\omega = 2\omega_p$ (see [31]).

J × B heating applies for S and P polarization, also at normal incidence, while for C-polarization and normal incidence it can be shown that this mechanism is totally suppressed.

The absence of efficient electron heating mechanisms for C-polarization at normal incidence is exploited in ion acceleration schemes which rely solely on the radiation pressure on the target. In these schemes indeed, the generation of hot electrons is usually detrimental, since it may lead to an early disruption of the target.

2.2.5 Target Normal Sheath Acceleration (TNSA)

In the previous subsection, it was mentioned that electron heating processes in solid density targets may lead to ion acceleration. Here, the most studied ion acceleration mechanism, TNSA (Target Normal Sheath Acceleration) is exposed in some detail. Other interesting ion acceleration schemes exist and they are briefly mentioned in Sect. 2.2.7. The interested reader is referred to [31, 36].

Ions are at least $\sim$2000 times heavier than electrons. Unless the laser pulse intensity exceeds $10^{24} - 10^{25}$ W/cm^2 (far beyond what can be achieved with modern laser technology), the electric field of the pulse cannot accelerate directly the ions.

However, intense laser pulses can transfer a significant fraction of their energy to the electrons of the target, whose expansion leads to strong electrostatic fields. Ions can be accelerated up to high energies ($\sim$60 MeV at the moment) by these fields. The observation of these high energy ions dates back to the early 2000 (these experiments where performed with high energy "long" ns pulses, only recently experiments performed with fs Ti:Sapphire lasers have matched these results).

When an intense laser pulse interacts with a solid-density plasma, it transfers a significant fraction of its energy to a hot electron population. When hot electrons escape from the rear side of the target, they produce a sheath field E_s normal to the target surface. The magnitude of the sheath field can be estimated as $E_s = \dfrac{T_h}{eL_s}$, where $T_h \sim m_e c^2(\sqrt{1+a_0^2/2}-1)$ is the hot electron temperature, e is the elementary charge and L_s is the scalelength of the sheath field. L_s can be roughly approximated as $L_s \sim L_D$, where $L_D = \sqrt{T_h/4\pi e^2 n_h}$ is the Debye length of the hot electrons.

Assuming a 10% laser absorption efficiency, a solid density plasma and a pulse intensity of $\sim 10^{20}$ W cm^2 the aforementioned estimations lead to a sheath field of $\sim 6 \times 10^{10}$ V/cm, which is orders of magnitude larger than what can be obtained in a conventional particle accelerator. Of course the sheath field decays rapidly after a few µm. Nonetheless, it is strong enough to ionize the impurities at the target surface (mainly carbon and hydrogen) and accelerate them up to a few tens of MeVs per nucleon. The acquired energy can be simply estimated as $E \sim Z e E_s L_s$, where Z is the atomic number of the ion. Replacing the formulae for E_s and L_s in the previous expression, we obtain a scaling law $E \sim I^{1/2}$, where I is the laser intensity. This is a rather unfavourable scaling, compared to other acceleration schemes like Radiation Pressure acceleration.

Reference [41] provides an extensive comparison of the theoretical models formulated for TNSA, concluding that the quasi-static approach [42] is particularly reliable to predict the energy of the accelerated ions.

2.2.6 Radiation Reaction Force and QED Effects

Laser-matter interaction with ultra-intense laser pulses leads to extreme electron acceleration, which become ultra-relativistic in a single laser-cycle. As widely known, an accelerated particle irradiates EM energy. The back-reaction force exerted on the particle due to this EM emission is called *Radiation Reaction* force (RR). Taking into account RR is important since it may significantly affect electron dynamics in laser matter interaction at intensities exceeding $I \sim 10^{23}$ W/cm^2, which will be available in the upcoming, next-generation laser facilities. Also the simulation of astrophysical scenarios involving ultra-relativistic plasmas may require RR to be considered.

RR can be considered as an additional term to the Lorentz fore:

$$\frac{d\mathbf{p}}{dt} = -e\left(\mathbf{E} + \frac{\mathbf{v}}{c} \times \mathbf{B}\right) + \mathbf{f}_{rad} \qquad (2.29)$$

Considering the power dissipated by an accelerated charged particle:

$$P_{rad} = \frac{2e^2}{3c^3}|\dot{\mathbf{v}}|^2 = \frac{2e^2\omega_c^2}{3c^3}v^2 \qquad (2.30)$$

we may naively write an expression for $\mathbf{f}_{rad}$, so that $\int_0^t \mathbf{f}_{rad} \cdot \mathbf{v} dt' = -\int_0^t P_{rad} dt'$. This approach leads to

2.2 Relativistic Laser Plasma Interaction

$$\mathbf{f}_{rad} = \frac{2e^2}{3c^3}\dddot{\mathbf{v}} \qquad (2.31)$$

which is however unsatisfactory, since it allows runaway solutions.

In [43], the following expression for RR is derived:

$$\mathbf{f}_{rad} = \frac{2r_c^2}{3}\left\{-\gamma^2\left[\left(\mathbf{E}+\frac{\mathbf{v}}{c}\times\mathbf{B}\right)^2-\left(\frac{\mathbf{v}}{c}\cdot\mathbf{E}\right)^2\right]\frac{\mathbf{v}}{c}+\right.$$
$$\left.+\left[\left(\mathbf{E}+\frac{\mathbf{v}}{c}\times\mathbf{B}\right)\times\mathbf{B}+\left(\frac{\mathbf{v}}{c}\cdot\mathbf{E}\right)\mathbf{E}\right]-\gamma\frac{m_ec}{e}\left(\dot{\mathbf{E}}+\frac{\mathbf{v}}{c}\times\dot{\mathbf{B}}\right)\right\} \qquad (2.32)$$

where $\dot{\mathbf{E}} = (\partial_t + \mathbf{v}\cdot\nabla)\mathbf{E}$ and $\dot{\mathbf{B}} = (\partial_t + \mathbf{v}\cdot\nabla)\mathbf{B}$. The expression for RR of Eq. 2.32 does not lead to runaway solutions.

A simple and effective numerical technique to include RR effects in PIC codes is presented in [44]. Despite its simplicity, this approach is able to reproduce the exact solution of Eq. 2.29 (obtained with expression 2.32 for f_{rad} [45]).

The description given for RR is completely classical. This means that this treatment is not valid when the laser field approaches the *Schwinger limit* in the reference frame co-moving with the electron:

$$E_s = \frac{m_ec^2}{\lambda_c} = 1.3\times 10^{16}\,\mathrm{V\,cm^{-1}} \qquad (2.33)$$

In this regime the electric field becomes high enough to generate electron-positron pairs from the vacuum. In the laboratory frame, the *Schwinger limit* corresponds to a laser intensity of $\sim 10^{29}$ W/cm^2 (for Ti:Saphphire lasers), which is extremely higher than what can be achieved with present and foreseen laser technology. However, in the reference frame co-moving with the electron, the electric field is up-shifted by a factor 2γ. This means that, in specific conditions, the onset of QED effects in laser-plasma interactions may be observed at much lower intensities [46].

2.2.7 Applications

In this section a condensed summary of the main trends in intense laser-plasma interaction research is given. An exhaustive treatment of these topics is far beyond the scope of this document. The following list is provided in order to account for the variety of this research field.

Electron Acceleration
Electron acceleration is a long-standing topic in intense laser-plasma interaction. The idea of exploiting a laser-induced wake in an under-dense plasma for electron acceleration dates back to 1979 [47]. This scheme is known as *Laser WakeField Acceleration (LWFA)*

For Ti:Sapphire laser systems, targets with $n < n_c$ are normally in gaseous form (gas jets synchronized with the laser pulse are used). Essentially, with a laser pulse propagating in a sub-critical plasma it is possible to induce periodic perturbations of the electron densities (this is due to the ponderomotive force). These density perturbations (see Fig. 2.6) are associated with a longitudinal electric field. Some electrons (coming from the plasma or externally injected) can be trapped inside the lower density region and, under the right conditions, they can be accelerated up to very high energies. Several different strategies have been explored in this field and in 2004 electron energies of $\sim$200 MeV (see [48]) were reached, while the current record is in the multi-GeV range (see [49]). The most attractive features of these plasma based accelerators are the extreme compactness of the source (the acceleration takes place in a few millimetres of plasma) and the very high intensity of the accelerated electron bunch (this is due to the fact that the bunch is extremely short in time, few 10s fs). Due to these properties, LWFA is also attractive as a secondary source of radiation.

A significant research effort in this field is currently focused on multi-stage LWFA accelerators [50], in which an electron bunch generated with LWFA is injected in another laser-driven wakefield. The final goal of this research line is achieving performances comparable with conventional accelerators (tens or hundreds of stages are though to be required for this purpose).

Recently, approaches able to exploit the foreseen advancements in laser technology have been proposed (e.g. multi-fibre lasers), like plasma wakefield accelerators driven by an incoherent combination of laser pulses [51].

It is worth to mention here that, though not based on a laser driven wakefield, plasma wakefield accelerators (PWFA) have been tested successfully to boost the energy of electron bunches accelerated by a conventional Linac (see [52] for a recent result). The main idea behind PWFA is very similar to LWFA, except that the wakefield is generated by the electron bunch itself.

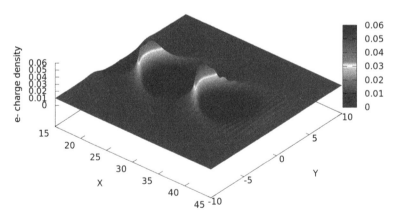

Fig. 2.6 Laser WakeField Acceleration. Laser induced electron density perturbations are clearly visible. Strong longitudinal electric fields are associated with these perturbations

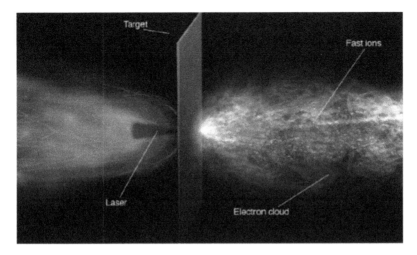

Fig. 2.7 Target Normal Sheath Acceleration principle. Reprinted figure with permission from A. Macchi, M. Borghesi, and M. Passoni, Reviews of Modern Physics 85, 751 [36]. Copyright 2013 by the American Physical Society

Ion Acceleration

Besides electrons, also ions can be accelerated by means of laser plasma interaction (see [36, 53] for a review). Normally, solid targets are used in this research field. The interaction of the laser pulse with solid targets results in the generation of a very high density plasma (typically $\sim 300\, n_c$).

Ions are too heavy to be accelerated directly by the laser pulse (even for the laser systems which will come online in the next years). Thus their acceleration is essentially due to the electrostatic field generated by the electrons. In a typical scenario, the so-called Target Normal Sheath Acceleration (TNSA), thin solid targets are used. The interaction of the laser pulse at the target surface generates a population of hot electrons. These electrons are expelled from the target, generating a very intense longitudinal electrostatic field at the back side. This electric field in turn accelerates the ions coming from the back side of the target[13] (see Fig. 2.7 for an artistic representation).

Several other ion acceleration schemes exist. Some are variations of the TNSA scheme which use structured targets to improve hot electron generation (coupled foam-solid targets or targets with a micro-structured surface). This topic will be covered extensively in Chap. 5.

Other schemes, such as Radiation Pressure Acceleration (RPA), differ significantly from the TNSA picture. In RPA, indeed, electron heating is suppressed and a charge separation is directly induced by the radiation pressure exerted on the electrons of the target (see Chap. 6).

Normally ion acceleration schemes are able to accelerate ions up to a few tens of MeV per nucleon. As for electron accelerators, laser-based ion acceleration schemes

---

[13] Normally a layer of hydrocarbon contaminants is always present on the target surfaces.

are interesting for their compactness. Moreover, if significant improvements over present laser plasma accelerators were achieved, they could become attractive for medical purposes. In fact, ions in the energy range of 100s MeV are routinely used for cancer treatment in several medical facilities in the world (the technique is called *hadron-therapy*). Ions are useful for cancer treatment because they release most of their energy at the Bragg peak. This means that it is possible to carefully select the ion energy in order to maximize the tissue damage on the cancer region, sparing the healthy tissues.

Medical facilities able to treat cancer with hadron-therapy use traditional accelerators (which are very expensive and require very large facilities). However, before plasma based accelerators could compete with traditional accelerators in this field, several important improvements should be achieved. In particular, laser-accelerated ions exhibit a cut-off energy which is still too low to be of medical interest. Moreover, their spectrum is extremely broad, while a quasi mono-energetic spectrum would be required. Finally, reliability of these plasma based systems should be improved.

Besides the medical use envisaged for laser-based ion sources, other interesting applications in material science may significantly relax the requirements for the accelerated ions. Indeed, present laser-based ion acceleration schemes may be already suitable or not too far from being suitable for some of these applications. This topic will be discussed in more detail in Chap. 5.

High Harmonic Generation

High Harmonic Generation (HHG) (see [54]) is a vibrant research field, stemmed from the study of laser-matter interaction. Irradiating a gaseous target with relatively intense ultra-short laser pulses results in the generation of high-order harmonics. The semi-classical *recollision* model [55] is helpful to get an insight of the process. In the intense laser field, an electron can tunnel through the atomic potential into the vacuum, where it can be accelerated by the laser. After a half optical cycle the laser is reversed and the electron can recombine with its parent atom, emitting a burst of radiation.

The most attractive feature of HHG is the possibility to tune the process to generate extremely short laser pulses, down to the attosecond timescale [56], which are very attractive as a diagnostic tool for ultra-fast electronic processes in atoms and molecules.

In order to generate high-order harmonics and then attosecond pulses with gaseous targets, the intensity of the laser pulse cannot be greater than $\sim 10^{16}$ W/cm^2, which prevents the generation of *high intensity* attosecond pulses (the maximum energy of an intense attosecond pulse is $\sim 1\,\mu$J [57]). A possible route to higher-intensity attosecond pulse sources consists in using overdense plasma surfaces and ultra-high intensity laser pulses [58, 59]. There are currently ongoing efforts to optimize harmonic generation with this scheme and to produce and isolate attosecond pulses. The topic will be covered more extensively in Chap. 6.

Intense Gamma Sources

Electrons accelerated via the LWFA process may be exploited as a secondary source of radiation. Schemes based on Thomson scattering may be implemented to obtain compact and intense x-rays/gamma sources (see for example [60]).

Recently, a laser-based γ source set the record for the highest peak brilliance in the multi-MeV energy range: $\sim 2 \times 10^{20}$ photons $s^{-1} mm^{-2} mrad^{-2}$ (see [61]). This value exceeds by several orders of magnitude the peak brilliance that can be obtained with conventional sources available in that energy range.

Laser-Based Neutron Sources

Laser-based particle acceleration (both of electrons and ions) opens interesting perspectives for compact, ultra-intense and ultra-short laser-based neutron sources.

Two mechanisms can be exploited: laser-based electron acceleration for photoneutron generation and nuclear reactions induced by laser-accelerated ions. As far as the first mechanism is of concern, Bremsstrahlung photons are generated with electrons in the in the multi-MeV or 10s Mev energy range interacting with a suitable solid target. The energy spectrum of these photons extends up to the initial electron energy [62] and it is thus suitable for photoneutron generation. Laser-based sources of photoneutrons were described in the past [63] and recently a few experiments have reported very high intensity neutron fluxes, exceeding several other bright neutron sources [64–66]. As far as the second mechanism is of concern, a "traditional" laser based ion acceleration scheme is coupled with a suitable target for neutron generation (see [66]).

As pointed out in [64], such short and intense pulsed neutron sources may have significant applications for the Fast Neutron Resonance Radiography technique [67] (which is useful for active material interrogation [68–71]). Moreover neutron sources based on laser systems foreseen in the near future may even allow to study nucleosynthesis processes in extreme astrophysical scenarios [72].

Laboratory Astrophysics

The availability of high intensity laser sources allows to test mechanisms relevant for astrophysics in a laboratory environment (the so-called *laboratory astrophysics*). This allows to study physical scenarios which otherwise would be beyond reach for direct experimental testing. Several questions concerning crucial astrophysical processes are indeed still open and the research effort on this topic may take advantage from experimental tests. Scaling the parameters of astrophysical plasmas down to the laboratory scale can be done rigorously (see [73]). Typical laboratory astrophysics experiments involve the study of shocks and/or counter-streaming plasmas (see [74]).

Recently the generation of neutral and high-density electron-positron pair plasmas in the laboratory was demonstrated [75] and numerical studies [76] indicate that upcoming laser facilities will be able to provide pair plasmas with parameters suitable to test astrophysical scenarios.

Finally, energetic ns beams were used in the past to compress matter much beyond what can be achieved with diamond anvils (see [77] for an example): NIF facility (USA) was recently able to reach a pressure of ~ 1 Gbar [78]. These studies are useful to understand the properties (crystalline structure, conductibility ...) of matter in conditions analogous to those found in planetary cores. Recent results with nanostructured targets [79] suggest that extremely high pressures suitable for High Energy Density physics studies can be achieved even with ultra-high intensity fs lasers.

Laser-Plasma Interaction at Extreme Laser Intensities
The advent of future laser facilities should allow to the observation of physical processes at energy scales beyond those of atomic physics (see [80] for a thorough review). For instance, laser intensities exceeding 10^{23} W/cm^2 should be high enough to observe an electron dynamics strongly dominated by RR. Even higher intensities may allow in the future to test QED phenomena such as QED cascades and vacuum polarization.

2.3 High Field Plasmonics

The excitation and control of Surface Plasmons (SPs) is a vibrant research field [81, 82], with several present and foreseen applications, ranging from extreme light concentration beyond diffraction limit [83] to biosensors [84] and plasmonic chips [85].

SPs are collective excitations of the electrons at the interface between a metal and a dielectric. "Classical" plasmonics schemes involve the interaction of low intensity ($I < 10^{12}$ W/cm^2) laser pulses with sub-wavelength structured targets (target structuring is required for SP excitation).

Some configurations of classical plasmonics may be extended into the high intensity laser-matter interaction regime. Plasmonics in this regime, where strongly nonlinear effects are expected, is essentially unexplored.[14] SP excitation at relativistic intensities poses new questions and might open new frontiers for manipulation and amplification of high power laser pulses.

Section 2.3.1 provides an introduction on SP physics with an eye towards high field regime. An overview of classical plasmonics schemes is provided in Sect. 2.3.2, while in Sect. 2.3.3 the outlook for the research on high field plasmonics is discussed.

2.3.1 Excitation of Surface Plasmons

In this section, only an introductory theoretical background on SP will be given. For a thorough treatment the interested reader is referred to [87, 88] (see instead [89] for a discussion on High Field Plasmonics).

SPs are electron oscillation modes which are confined at a steep interface between a metal and a dielectric material. We will first derive their dispersion relation before discussing their properties and how they can be excited.

---

[14]The excitation of surface waves in relativistic laser-matter interaction was first proposed in [86] but the topic has remained essentially unexplored up to now.

2.3 High Field Plasmonics

In a wide range of frequencies (far from the absorption regions and neglecting damping), the dielectric function of a metal can be approximated with the dielectric function of a plasma: $\epsilon_r(\omega) = 1 - \omega_p^2/\omega^2$, where ω_p is the plasma frequency. We will consider a perfectly flat interface between a metal ($\epsilon_2 = \epsilon_r(\omega)$) and a generic dielectric medium, characterized by the dielectric constant ϵ_1. We will adopt the reference frame of Fig. 2.8: $\hat{\mathbf{x}}$ is perpendicular to the surface, while $\hat{\mathbf{y}}$ is the direction of propagation along the interface. The system is taken to be invariant along $\hat{\mathbf{z}}$. Throughout the derivation subscripts $_1$ and $_2$ will be used to indicate, respectively, a field in the dielectric region ($x > 0$) or in the metal region ($x < 0$).

Given the system of Fig. 2.8, the following conditions for the EM field should be imposed:

$$\begin{cases} \mathbf{E}_{\|1} = \mathbf{E}_{\|2} \\ \mathbf{B}_{\|1} = \mathbf{B}_{\|2} \\ \epsilon_1 \mathbf{E}_{\perp 1} = \epsilon_2 \mathbf{E}_{\perp 2} \\ \mathbf{B}_{\perp 1} = \mathbf{B}_{\perp 2} \end{cases} \quad (2.34)$$

Subscripts $\perp$ and $\|$ indicate, respectively, the field component perpendicular to the surface and the field components parallel to the surface. It can be shown that no transverse electric modes (TE) can exist with these conditions (see [88]). We thus consider a transverse magnetic (TM) mode, for which we make the following ansatz (monochromatic SP):

$$\begin{cases} \mathbf{B} = B_0 \hat{\mathbf{z}} e^{-qx} e^{i(ky - \omega t)} \\ \mathbf{E} = (E_{0x} \hat{\mathbf{x}} + E_{0y} \hat{\mathbf{y}}) e^{-qx} e^{i(ky - \omega t)} \end{cases} \quad (2.35)$$

where q may be different in the two regions.

Using $\nabla \times \mathbf{B} = \mu \left(\dfrac{4\pi}{c} \mathbf{J} + \dfrac{\epsilon}{c} \dfrac{\partial \mathbf{E}}{\partial t} \right)$ and performing a Fourier transform with respect to time we get $\nabla \times \tilde{\mathbf{B}} = -i\omega\epsilon\tilde{\mathbf{E}}/c$ (no current, magnetic permeability $\mu \sim 1$).

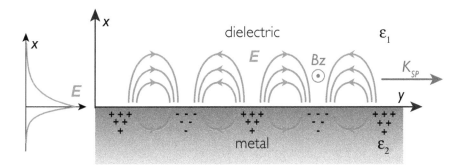

Fig. 2.8 Schematic representation of the surface oscillation mode of the electrons at the steep metal-dielectric interface. The figure is reproduced with permission from [89]

If we insert the ansatz in Eq. 2.35 we finally get:

$$\tilde{\mathbf{E}} = \frac{-c}{\omega\epsilon}\left[k\hat{\mathbf{x}} - iq\hat{\mathbf{y}}\right] B_0 \tag{2.36}$$

Imposing the boundary conditions of Eq. 2.34 we find:

$$\frac{q_1}{\epsilon_1} = \frac{q_2}{\epsilon_2} \tag{2.37}$$

Since we are looking for solutions bounded to the interface, q_1 and q_2 should be real and with opposite sign (q_1 should be positive, q_2 should be negative in order to avoid divergences at $x = \pm\infty$). This is only possible if ϵ_1 and ϵ_2 have opposite sign (i.e. if one medium is a dielectric and the other one is a metal).

In general, in a medium with dielectric function $\epsilon(\omega)$, the magnetic field $\mathbf{B}$ satisfies Helmholtz equation $\left(\nabla^2 + \epsilon\frac{\omega^2}{c^2}\right)\mathbf{B} = 0$. This implies:

$$q^2 - k^2 + \frac{\omega^2}{c^2}\epsilon = 0 \tag{2.38}$$

Since k should be the same for the metal region and the dielectric region:

$$q_1^2 + \frac{\omega^2}{c^2}\epsilon_1 = q_2^2 + \frac{\omega^2}{c^2}\epsilon_2 \tag{2.39}$$

using Eq. 2.37 in 2.39 we get $q_1^2 = -\epsilon_1^2\frac{\omega^2}{c^2}\frac{1}{\epsilon_1+\epsilon_2}$. Plugging this last result into 2.38 and solving in region 1, we finally get the dispersion relation for a SP:

$$k(\omega) = \frac{\omega}{c}\sqrt{\frac{\epsilon_1\epsilon_2}{\epsilon_1+\epsilon_2}} \tag{2.40}$$

For the specific case of an interface between a metal and the vacuum (in which we will mainly interested) Eq. 2.40 becomes:

$$k(\omega) = \frac{\omega}{c}\sqrt{\frac{1-\omega_p^2/\omega^2}{2-\omega_p^2/\omega^2}} \tag{2.41}$$

Figure 2.8 shows the propagation of a SP. The upper material is the dielectric, while the lower material is a metal. It is worth to remark that the electromagnetic field extends very little into the metal. In fact, the decay length is essentially the electron skin depth in this case.

The dispersion relation of Eq. 2.41 has an evident singularity when $\omega \to \omega_p/\sqrt{2}$. Then there's a gap for $\omega_p/\sqrt{2} < \omega < \omega_p$ where $k(\omega)$ is imaginary and finally there's

2.3 High Field Plasmonics

another branch of the dispersion function for $\omega > \omega_p$. For SPs only the lower branch is of interest $\left(\omega < \omega_p/\sqrt{2}\right)$, since the upper branch describes the propagation of very high frequency EM waves in the metal.

The previous derivation was performed for a metal. However it is valid also for a cold plasma, provided that relativistic effects are disregarded.

When $\omega < \omega_p/\sqrt{2}$, $k(\omega) > \omega/c$. This implies that coupling between an EM wave in vacuum and a SP is impossible. In fact, the matching condition for an EM wave with an angle of incidence θ reads as follow:

$$k_{EM}(\omega)\sin(\theta) = k_{SPP}(\omega) \longrightarrow \frac{\omega}{c}\sin(\theta) = k_{SP}(\omega) \quad (2.42)$$

Equation 2.42 cannot be satisfied if $\omega < \omega_p$ (in the previous equation k_{EM} and k_{SP} are, respectively, the k vector for the EM wave and the SP).

In general, the excitation of a surface plasmon by direct illumination of a flat interface can be proven to be impossible. Thus several schemes have been developed in order to excite a SP with an electromagnetic wave.[15] In the so-called *Otto* configuration, a dielectric prism is in direct contact with a metal thin surface. When an EM field is reflected at the dielectric-metal interface, its evanescent field component is responsible for the excitation of a surface plasmon on the back side of the metal thin foil. *Kretschmann* configuration is a very similar scheme, but it this case the aforementioned components are arranged is such a way to introduce a thin air gap between the prism and the metal. In this case the SP is excited by the evanescent EM field on the face of the metal foil facing the prism. These strategies are crucially based on the use of a dielectric.

The aforementioned schemes are clearly unsuitable for the aim of extending the study of plasmonics effects in high intensity laser-matter interaction. Indeed during the interaction with a high intensity laser pulse any dielectric becomes a strongly conductive plasma. Therefore, we must rely on other common schemes in "traditional" plasmonics involving a structuring of a metal interface. Among these techniques a widely used scheme consists in using metal surfaces with a shallow modulation of period d (if the modulation of the surface is not shallow, the dispersion relation of the SP can be strongly affected). The SP dispersion relation is then folded with period $q = 2\pi/d$ in the $k - \omega$ plane and the matching condition hence reads:

---

[15] Without using an EM wave, a SP can be excited also with accelerated charged particles.

$$K_{EM}(\omega)\sin(\theta) = k_{SPP}(\omega) \pm n\frac{2\pi}{d} \quad (2.43)$$

where d is the groove spacing and n is an integer number.

A solution for the matching condition of Eq. 2.43 can then be found. For a given value of ω, the condition of Eq. 2.43 depends only on the period of the grating d and the angle of incidence of the EM wave ϕ_i:

$$\pm n\lambda_L/d = \sqrt{(1-\omega_p^2/\omega^2)/(2-\omega_p^2/\omega^2)} - \sin(\phi_i). \quad (2.44)$$

For a strongly over-dense plasma ($\omega_p \gg \omega$), if we restrict to the solution with $n = +1$, the conditions for d and ϕ_i becomes:

$$\frac{\lambda_L}{d} = 1 - \sin(\phi_i). \quad (2.45)$$

The matching condition is represented graphically in Fig. 2.9. Equation 2.45 is plotted in Fig. 2.10. Some selected values are highlighted.

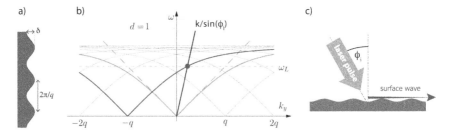

Fig. 2.9 Pictorial representation of the excitation of a SP by a plane wave shining on a metal interface. The figure is reproduced with permission from [89]

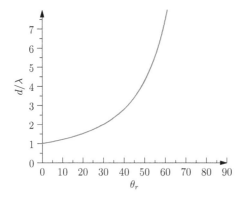

θ_r	d/λ
0	1
10°	1.210
20°	1.520
30°	2
45°	3.414

Fig. 2.10 Relation between grating resonance angle and groove spacing d

2.3 High Field Plasmonics

The theory presented so far is valid for non-relativistic laser-plasma interaction. It is important to stress that we are mainly interested in the extension of Plasmonics in the High Field regime, where we expect relativistic effects to be strong. Thus there's no "a priori" guarantee that the theory which we've exposed will hold at relativistic intensities. Indeed, a complete and convincing theory for surface plasmon polaritons in the relativistic regime (High Field Plasmonics) is still lacking in the literature.

2.3.2 Overview of Plasmonic Schemes and Applications

The impressive development of plasmonics in the last decades has led to the realization of a large amount of interesting schemes. Figure 2.11 gives an idea of the variety of Plasmonics schemes and its applications.

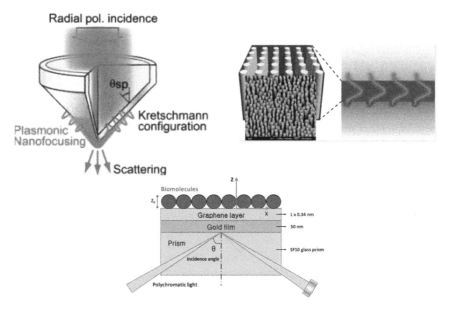

Fig. 2.11 A few selected applications of plasmonics, which illustrate the variety of the field. Picture 1 shows a nanofocusing device, picture 2 shows a plasmonic metamaterial made of metallic nanorods, picture 3 shows the use of a plamonic device for biosensing: the high field confinement achieved in plasmonic schemes is exploited to enhance biomolecules detection. Picture 1 is reproduced from Ono et al. Surface Plasmon Excitation and Localization by Metal-Coated Axicon Prism. Micromachines,3 (2012) [90]. Copyright 2012 by the authors, released under a CC-BY license http://creativecommons.org/licenses/by/3.0/. Picture 2 is reproduced from Vasilantonakis et al. Laser & Photonics Reviews,9 (2015) [91]. Copyright 2015 by the authors, released under a CC-BY license http://creativecommons.org/licenses/by/3.0/. Picture 3 is reproduced from Nguyen et al. Surface plasmon resonance: a versatile technique for biosensor applications. Sensors, 15 (2015) [92]. Copyright by the authors, released under a Creative Commons attribution license http://creativecommons.org/licenses/by/4.0/

In the the left upper panel, a device conceived to achieve plasmonic nanofocusing is shown. While propagating EM waves cannot be focused below the diffraction limit $\lambda/2$, surface plasmons are evanescent modes and can be concentrated much further. A variety of devices able to concentrate surface plasmons in structures much smaller than the wavelength of the exciting EM have been developed [93] (nanotips, tapered waveguides, kissing cylinders …) leading to extreme field enhancements.[16] Plasmonic nanofocusing in metallic tips is currently exploited for technological applications such as the Scanning Near-field Optical Microscopy (SNOM) technique [94].

Plasmonic metamaterials [95] (see the top right panel of Fig. 2.11) are materials structured on a sub-wavelength scale which exploits plasmonics effects to achieve peculiar optical properties, which are not found in conventional materials. For instance, a plasmonic metamaterial may behave as a negative refraction index material. Suitably prepared plasmonics meta-surfaces may show unconventional optical properties as well (see [96]).

Plasmonic elements can be integrated into microfluidic lab-on-chips. The replacement of traditional electronic components with plasmonics equivalents is also an active research field [85].

An important technological application of SP is represented by biosensors (see [97]): a surface plasmon in excited on a suitably coated surface and the analysis of the reflected light can be used to gather information on the adsorbed molecules. The bottom panel of Fig. 2.11 provides a schematic of this technique.

Finally, the light trapping properties of SPs have been exploited to enhance the efficiency of solar cells [98].

Provided that plasmonics can be extended into the relativistic regime, the large variety of schemes already developed in traditional plasmonics could represent an interesting source of inspiration.

2.3.3 Outlook for Relativistic Plasmonics

The perspective of extending some schemes borrowed from classical plasmonics into the high field regime is certainly interesting. For instance, energy concentration schemes are particularly attractive, since they may lead to extreme laser intensities, currently out of reach for modern laser systems. Moreover, fine manipulation of ultrahigh fields with structured targets may open the way for interesting possibilities.

Though potentially far-reaching, the advent of High Field Plasmonics could be hindered by fundamental difficulties inherent to high intensity laser-matter interaction. Indeed, several plasmonics schemes require sub-wavelength structuring of the targets: shallow gratings are required for laser coupling to surface plasmons, subwavelength tapered waveguides (or simlar subwavelength structures) are required for light concentration beyond diffraction limit, subwavelength holes are required for plasmonic transparency schemes…However, in laser-matter interaction at relativistic

---

[16]Field enhancement exceeding $100\times$ are reported in the literature.

intensities, electrons reach almost the speed of light in one laser cycle. This means that they can be displaced by $\approx \lambda_{Laser}$ in a single laser cycle, which raises the concern that the subwavelength structures may not survive long enough.

Despite this issue, a few promising experimental results have been obtained recently with grating targets irradiated at relativistic intensities (see Chap. 4 for an extensive treatment). For instance, in a first experimental campaign in 2012 at CEA-Saclay, grating targets were irradiated with intense ($I > 10^{19}$ W/cm^2) laser pulses at the resonant angle for SP excitation. This configuration was proven to enhance the cut-off energy of protons accelerated from the rear surface of the target [99]. A second experimental campaign was performed at the same facility in 2014 [100, 101]; the activity was focused on the effects of SP excitation on electron emission from grating targets. A strong emission of multi-MeV electron bunches was observed along the target surface only when a grating target was irradiated with an angle close to the one expected for the resonant SP excitation. Especially this second experimental campaign provides compelling evidence of SP excitation (a theoretical model together with 3D numerical simulations confirm that SP are involved in the electron acceleration process). These scheme represents a potentially attractive electron source for a few applications (see Chap. 4).

In addition, plasmonics effects can be identified in the Light Sail regime of Radiation pressure ion acceleration scheme (see Chap. 6 for a detailed discussion). Experimental evidence of complex structures in RPA ions was found in a few experiments [102], suggesting that the RPA scenario is prone to surface rippling instabilities, and structures closely resembling those determined by Rayleigh Taylor Instability were observed in several numerical studies. However, the scalelength of the structures observed in the simulations (close to the laser wavelength) cannot be explained with classical Rayleigh Taylor Instability theory. It is possible to show (both theoretically and numerically) that the plasmonic resonant coupling of the laser light with the target rippling affects the growth of RTI driven by the radiation pressure, determining the scale of the target rippling. These results are of interest because target rippling is possibly detrimental for the RPA scheme (since it may lead to an early onset of transparency).

In conclusion, though probably not all the schemes from traditional plasmonics can be easily ported into the high field regime, experimental and numerical evidence strongly supports SP excitation in relativistic laser-matter interaction. Besides their specific interest, these results open the way for further developments and support High Field Plasmonics as an emerging field in ultra-high intensity laser-matter interaction.

References

1. J.P. Gordon, H.J. Zeiger, C.H. Townes, The maser new type of microwave amplifier, frequency standard, and spectrometer. Phys. Rev. **99**, 1264–1274 (1955)
2. O. Svelto, D.C. Hanna, *Principles of Lasers* (Springer, Berlin, 2009)
3. T.H. Maiman, Stimulated optical radiation in ruby. Nature **187**, 493–494 (1960)
4. (editorial). Fifty brilliant years. Nat. Mater. (2010)

5. V. Yanovsky, V. Chvykov, G. Kalinchenko, P. Rousseau, T. Planchon, T. Matsuoka, A. Maksimchuk, J. Nees, G. Cheriaux, G. Mourou, K. Krushelnick, Ultra-high intensity- 300-TW laser at 0.1 Hz repetition rate. Opt. Express **16**(3), 2109–2114 (2008)
6. Wikipedia (picture released as "public domain"). History of laser intensity (2007); (online). Accessed 01 May 2016
7. F.J. McClung, R.W. Hellwarth, Giant optical pulsations from ruby. J. Appl. Phys. **33**(3), 828–829 (1962)
8. A.H. Haus, Mode-locking of lasers. IEEE J. Sel. Top. Quantum Electron. **6**(6), 1173–1185 (2000)
9. International Energy Agency (ed.), *Key World Energy Statistics* (IEA, 2014)
10. J.E. Geusic, H.M. Marcos, L.G. Van Uitert, Laser oscillations in Nd-doped yttrium aluminium, yttrium gallium and gadolinium garnets. Appl. Phys. Lett. **4**(10), 182 (1964)
11. C.K.N. Patel, Continuous-wave laser action on vibrational-rotational transitions of CO_2. Phys. Rev. **136**, A1187–A1193 (1964)
12. I.V. Pogorelsky, I. Ben-Zvi, J. Skaritka, M. Babzien, M.N. Polyanskiy, Z. Najmudin, N. Dover, W. Lu, New opportunities for strong-field LPI research in the mid-IR. Proc. SPIE **9509**, 95090P–95090P-10 (2015)
13. M.N. Polyanskiy, M. Babzien, I. Pogorelsky, V. Yakimenko. Ultrashort-pulse CO_2 lasers: ready for the race to petawatt? Proc. SPIE **8677**, 86770G–86770G-6 (2013)
14. D. Strickland, G. Mourou, Compression of amplified chirped optical pulses. Opt. Commun. **55**(6), 447–449 (1985)
15. Wikipedia (figure released as "public domain"). Chirped pulse amplification (2006); (online). Accessed 01 May 2016
16. B. Dromey, S. Kar, M. Zepf, P. Foster, The plasma mirror - a subpicosecond optical switch for ultrahigh power lasers. Rev. Sci. Instrum. **75**(3), 645–649 (2004)
17. C. Thaury, F. Quéré, J.-P. Geindre, A. Levy, T. Ceccotti, P. Monot, M. Bougeard, F. Reau, P. d'Oliveira, P. Audebert, R. Marjoribanks, Ph. Martin, Plasma mirrors for ultrahigh-intensity optics. Nat. Phys. **3**, 424–429 (2007)
18. C. Danson, D. Hillier, N. Hopps, D. Neely, Petawatt class lasers worldwide. High Power Laser Sci. Eng. **3** (2015)
19. I.W. Choi, Upgrade to 4 PW of the PULSER laser system, CoReLS, Institute for Basic Science, GIST, Gwangju, South Korea. personal communication
20. G. Chériaux, F. Giambruno, A. Fréneaux, F. Leconte, L.P. Ramirez, P. Georges, F. Druon, D.N. Papadopoulos, A. Pellegrina, C. Le Blanc, I. Doyen, L. Legat, J.M. Boudenne, G. Mennerat, P. Audebert, G. Mourou, F. Mathieu, J.P. Chambaret, Apollon-10P: status and implementation. AIP Conf. Proc. **1462**(1), 78–83 (2012)
21. C. Hernandez-Gomez, S.P. Blake, O. Chekhlov, R.J. Clarke, A.M. Dunne, M. Galimberti, S. Hancock, R. Heathcote, P. Holligan, A. Lyachev, P. Matousek, I.O. Musgrave, D. Neely, P.A. Norreys, I. Ross, Y. Tang, T.B. Winstone, B.E. Wyborn, J. Collier, The Vulcan 10 PW project. J. Phys.: Conf. Ser. **244**(3), 032006 (2010)
22. A. Dubietis, G. Jonušauskas, A. Piskarskas, Powerful femtosecond pulse generation by chirped and stretched pulse parametric amplification in BBO crystal. Opt. Commun. **88**(4–6), 437–440 (1992)
23. I.N. Ross, P. Matousek, M. Towrie, A.J. Langley, J.L. Collier, The prospects for ultrashort pulse duration and ultrahigh intensity using optical parametric chirped pulse amplifiers. Opt. Commun. **144**(1–3), 125–133 (1997)
24. I.N. Ross, P. Matousek, G.H.C. New, K. Osvay, Analysis and optimization of optical parametric chirped pulse amplification. J. Opt. Soc. Am. B **19**(12), 2945–2956 (2002)
25. X. Liang, L. Yu, L. Xu, Y. Chu, Y. Xu, C. Wang, X. Lu, Y. Leng, R. Li, Z. Xu. Latest progress and research status of ultra-high intensity lasers at siom, in *Advanced Solid State Lasers* (Optical Society of America, 2014), p. AM1A.1
26. T. Ebisuzaki, M.N. Quinn, S. Wada, L.W. Piotrowski, Y. Takizawa, M. Casolino, M.E. Bertaina, P. Gorodetzky, E. Parizot, T. Tajima, R. Soulard, G. Mourou, Demonstration designs for the remediation of space debris from the international space station. Acta Astronaut. **112**, 102–113 (2015)

27. V.Yu. Bychenkov, A.V. Brantov, G. Mourou, Tc-99m production with ultrashort intense laser pulses. Laser Particle Beams **32**, 605–611 (2014)
28. M. Gerard, B. Brocklesby, T. Tajima, J.J. Limpert, The future is fibre accelerators. Nat. Photonics **7**, 258–261 (2013)
29. C. Labaune, D. Hulin, A. Galvanauskas, G.A. Mourou, On the feasibility of a fiber-based inertial fusion laser driver. Opt. Commun. **281**(15–16), 4075–4080 (2008)
30. P. Gibbon, *Short Pulse Laser Interactions with Matter* (Imperial College Press, London, 2005)
31. A. Macchi, *A Superintense Laser-Plasma Interaction Theory Primer* (Springer, Netherlands, 2013)
32. P. Mulser, D. Bauer, *High Power Laser-Matter Interaction*, vol. 238 (Springer, Berlin, 2010)
33. S.R. Groot, W.A. Leeuwen, C.G. van Weert, *Relativistic Kinetic Theory: Principles and Applications* (North-Holland, Amsterdam, 1980)
34. A.W. Trivelpiece, N.A. Krall, *Principles of Plasma Physics* (McGraw-Hill, Englewood Cliffs, 1973)
35. J.D. Jackson, J.D. Jackson, *Classical Electrodynamics*, vol. 3 (Wiley, New York, 1962)
36. A. Macchi, M. Borghesi, M. Passoni, Ion acceleration by superintense laser-plasma interaction. Rev. Mod. Phys. **85**, 751–793 (2013)
37. F. Brunel, Anomalous absorption of high intensity subpicosecond laser pulses. Phys. Fluids (1958–1988) **31**(9), 2714–2719 (1988)
38. H. Popescu, S.D. Baton, F. Amiranoff, C. Rousseaux, M. Rabec Le Gloahec, J.J. Santos, L. Gremillet, M. Koenig, E. Martinolli, T. Hall, J.C. Adam, A. Heron, D. Batani, Subfemtosecond, coherent, relativistic, and ballistic electron bunches generated at ω_0 and $2\omega_0$ in high intensity laser-matter interaction. Phys. Plasmas **12**(6), 063106 (2005)
39. P. Mulser, H. Ruhl, J. Steinmetz, Routes to irreversibility in collective laser-matter interaction. Laser Particle Beams **19**, 23–28 (2001)
40. W.L. Kruer, K. Estabrook, J×B heating by very intense laser light. Phys. Fluids **28**(1), 430–432 (1985)
41. C. Perego, A. Zani, D. Batani, M. Passoni, Extensive comparison among target normal sheath acceleration theoretical models. Nucl. Instrum. Methods Phys. Res. Sect. A: Accel. Spectrom. Detect. Assoc. Equip. **653**(1), 89–93 (2011); Superstrong 2010
42. M. Lontano, M. Passoni, Electrostatic field distribution at the sharp interface between high density matter and vacuum. Phys. Plasmas **13**(4), 042102 (2006)
43. L.D. Landau, E.M. Lifshitz, *Course of Theoretical Physics Series (Book 2), vol. 2 (The Classical Theory of Fields), chapter 76* (Butterworth-Heinemann, London, 1980)
44. M. Tamburini, F. Pegoraro, A. Di Piazza, C.H. Keitel, A. Macchi, Radiation reaction effects on radiation pressure acceleration. New J. Phys. **12**(12), 123005 (2010)
45. A. Di Piazza, Exact solution of the Landau-Lifshitz equation in a plane wave. Lett. Math. Phys. **83**(3), 305–313 (2008)
46. A. Di Piazza, K.Z. Hatsagortsyan, C.H. Keitel, Quantum radiation reaction effects in multiphoton compton scattering. Phys. Rev. Lett. **105**, 220403 (2010)
47. T. Tajima, J.M. Dawson, Laser electron accelerator. Phys. Rev. Lett. **43**, 267–270 (1979)
48. J. Faure, Y. Glinec, A. Pukhov, S. Kiselev, S. Gordienko, E. Lefebvre, J.-P. Rousseau, F. Burgy, V. Malka, A laser-plasma accelerator producing monoenergetic electron beams. Nature **431**, 541–544 (2004)
49. W.P. Leemans, A.J. Gonsalves, H.-S. Mao, K. Nakamura, C. Benedetti, C.B. Schroeder, Cs. Tóth, J. Daniels, D.E. Mittelberger, S.S. Bulanov, J.-L. Vay, C.G.R. Geddes, E. Esarey, Multi-GeV electron beams from capillary-discharge-guided subpetawatt laser pulses in the self-trapping regime. Phys. Rev. Lett. **113**, 245002 (2014)
50. H.T. Kim, K.H. Pae, H.J. Cha, I.J. Kim, T.J. Yu, J.H. Sung, S.K. Lee, T.M. Jeong, J. Lee, Enhancement of electron energy to the multi-GeV regime by a dual-stage laser-wakefield accelerator pumped by petawatt laser pulses. Phys. Rev. Lett. **111**, 165002 (2013)
51. C. Benedetti, C.B. Schroeder, E. Esarey, W.P. Leemans, Plasma wakefields driven by an incoherent combination of laser pulses: a path towards high-average power laser-plasma accelerators). Phys. Plasmas **21**(5), 056706 (2014)

52. S. Corde, E. Adli, J.M. Allen, W. An, C.I. Clarke, C.E. Clayton, J.P. Delahaye, J. Frederico, S. Gessner, S.Z. Green, M.J. Hogan, C. Joshi, N. Lipkowitz, M. Litos, W. Lu, K.A. Marsh, W.B. Mori, M. Schmeltz, N. Vafaei-Najafabadi, D. Walz, V. Yakimenko, G. Yocky, Multi-gigaelectronvolt acceleration of positrons in a self-loaded plasma wakefield. Nature **524**, 442–445 (2015)
53. H. Daido, M. Nishiuchi, A.S. Pirozhkov, Review of laser-driven ion sources and their applications. Rep. Prog. Phys. **75**(5), 056401 (2012)
54. M. Lewenstein, Ph Balcou, MYu. Ivanov, A. L'Huillier, P.B. Corkum, Theory of high-harmonic generation by low-frequency laser fields. Phys. Rev. A **49**, 2117–2132 (1994)
55. P.B. Corkum, Plasma perspective on strong field multiphoton ionization. Phys. Rev. Lett. **71**, 1994–1997 (1993)
56. G. Sansone, E. Benedetti, F. Calegari, C. Vozzi, L. Avaldi, R. Flammini, L. Poletto, P. Villoresi, C. Altucci, R. Velotta, S. Stagira, S. De Silvestri, M. Nisoli, Isolated single-cycle attosecond pulses. Science **314**(5798), 443–446 (2006)
57. E.J. Takahashi, P. Lan, T. Okino, Y. Furukawa, Y. Nabekawa, K. Yamanouchi, K. Midorikawa, Intense attosecond pulses for probing ultrafast molecular dynamics, in *Ultrafast Phenomena XIX, vol. 162 of Springer Proceedings in Physics*, ed. by K. Yamanouchi, S. Cundiff, R. de Vivie-Riedle, M. Kuwata-Gonokami, L. DiMauro (Springer, Berlin, 2015), pp. 16–19
58. U. Teubner, P. Gibbon, High-order harmonics from laser-irradiated plasma surfaces. Rev. Mod. Phys. **81**, 445–479 (2009)
59. F. Quéré, C. Thaury, P. Monot, S. Dobosz, J.-P. Ph Martin, P.Audebert Geindre, Coherent wake emission of high-order harmonics from overdense plasmas. Phys. Rev. Lett. **96**, 125004 (2006)
60. S. Corde, K. Ta, Phuoc, G. Lambert, R. Fitour, V. Malka, A. Rousse, A. Beck, E. Lefebvre, Femtosecond x rays from laser-plasma accelerators. Rev. Mod. Phys. **85**, 1–48 (2013)
61. G. Sarri, D.J. Corvan, W. Schumaker, J.M. Cole, A. Di Piazza, H. Ahmed, C. Harvey, C.H. Keitel, K. Krushelnick, S.P.D. Mangles, Z. Najmudin, D. Symes, A.G.R. Thomas, M. Yeung, Z. Zhao, M. Zepf, Ultrahigh brilliance multi-MeV γ-ray beams from nonlinear relativistic Thomson scattering. Phys. Rev. Lett. **113**, 224801 (2014)
62. H.W. Koch, J.W. Motz, Bremsstrahlung cross-section formulas and related data. Rev. Mod. Phys. **31**, 920–955 (1959)
63. H. Schwoerer, P. Gibbon, S. Düsterer, R. Behrens, C. Ziener, C. Reich, R. Sauerbrey, MeV x-rays and photoneutrons from femtosecond laser-produced plasmas. Phys. Rev. Lett. **86**, 2317–2320 (2001)
64. I. Pomerantz, E. McCary, A.R. Meadows, A. Arefiev, A.C. Bernstein, C. Chester, J. Cortez, M.E. Donovan, G. Dyer, E.W. Gaul, D. Hamilton, D. Kuk, A.C. Lestrade, C. Wang, T. Ditmire, B.M. Hegelich, Ultrashort pulsed neutron source. Phys. Rev. Lett. **113**, 184801 (2014)
65. Y. Arikawa, M. Utsugi, A. Morace, T. Nagai, Y. Abe, S. Kojima, S. Sakata, H. Inoue, S. Fujioka, Z. Zhang, H. Chen, J. Park, J. Williams, T. Morita, Y. Sakawa, Y. Nakata, J. Kawanaka, T. Jitsuno, N. Sarukura, N. Miyanaga, H. Azechi, High-intensity neutron generation via laser-driven photonuclear reaction. Plasma Fusion Res. **10**, 2404003 (2015)
66. M. Roth, D. Jung, K. Falk, N. Guler, O. Deppert, M. Devlin, A. Favalli, J. Fernandez, D. Gautier, M. Geissel, R. Haight, C.E. Hamilton, B.M. Hegelich, R.P. Johnson, F. Merrill, G. Schaumann, K. Schoenberg, M. Schollmeier, T. Shimada, T. Taddeucci, J.L. Tybo, F. Wagner, S.A. Wender, C.H. Wilde, G.A. Wurden, Bright laser-driven neutron source based on the relativistic transparency of solids. Phys. Rev. Lett. **110**, 044802 (2013)
67. C. Gongyin, R.C. Lanza, Fast neutron resonance radiography for elemental imaging: theory and applications. IEEE Trans. Nucl. Sci. **49**(4), 1919–1924 (2002)
68. C.L. Fink, B.J. Micklich, T.J. Yule, P. Humm, L. Sagalovsky, M.M. Martin, Evaluation of neutron techniques for illicit substance detection. Nucl. Instrum. Methods Phys. Res. Sect. B: Beam Interact. Mater. Atoms **99**(1–4), 748–752 (1995); Application of Accelerators in Research and Industry'94
69. J. Rynes, J. Bendahan, T. Gozani, R. Loveman, J. Stevenson, C. Bell, Gamma-ray and neutron radiography as part of a pulsed fast neutron analysis inspection system. Nucl. Instrum.

Methods Phys. Res. Sect. A: Accel. Spectrom. Detect. Assoc. Equip. **422**(1–3), 895–899 (1999)
70. J.C. Overley, M.S. Chmelik, R.J. Rasmussen, R.M.S. Schofield, H.W. Lefevre, Explosives detection through fast-neutron time-of-flight attenuation measurements. Nucl. Instrum. Methods Phys. Res. Sect. B: Beam Interact. Mater. Atoms **99**(1–4), 728–732 (1995). Application of Accelerators in Research and Industry'94
71. J.E. Eberhardt, S. Rainey, R.J. Stevens, B.D. Sowerby, J.R. Tickner, Fast neutron radiography scanner for the detection of contraband in air cargo containers. Appl. Radiat. Isot. **63**(2), 179–188 (2005)
72. C. Freiburghaus, S. Rosswog, F.-K. Thielemann, r-process in neutron star mergers. Astrophys. J. Lett. **525**(2), L121 (1999)
73. S. Bouquet, E. Falize, C. Michaut, C.D. Gregory, B. Loupias, T. Vinci, M. Koenig, From lasers to the universe: scaling laws in laboratory astrophysics. High Energy Density Phys. **6**(4), 368–380 (2010)
74. É Falize, A. Ravasio, B. Loupias, A. Dizière, C.D. Gregory, C. Michaut, C. Busschaert, C. Cavet, M. Koenig, High-energy density laboratory astrophysics studies of accretion shocks in magnetic cataclysmic variables. High Energy Density Phys. **8**(1), 1–4 (2012). cited By 7
75. G. Sarri, K. Poder, J.M. Cole, W. Schumaker, A. Di Piazza, B. Reville, T. Dzelzainis, D. Doria, L.A. Gizzi, G. Grittani, S. Kar, C.H. Keitel, K. Krushelnick, S. Kuschel, S.P.D. Mangles, Z. Najmudin, N. Shukla, L.O. Silva, D. Symes, A.G.R. Thomas, M. Vargas, J. Vieira, M. Zepf, Generation of neutral and high-density electron–positron pair plasmas in the laboratory. Nat. Commun. **6**, 6747 (2015)
76. H. Chen, F. Fiuza, A. Link, A. Hazi, M. Hill, D. Hoarty, S. James, S. Kerr, D.D. Meyerhofer, J. Myatt, J. Park, Y. Sentoku, G.J. Williams, Scaling the yield of laser-driven electron-positron jets to laboratory astrophysical applications. Phys. Rev. Lett. **114**, 215001 (2015)
77. D. Batani, A. Morelli, M. Tomasini, A. Benuzzi-Mounaix, F. Philippe, M. Koenig, B. Marchet, I. Masclet, M. Rabec, Ch. Reverdin, R. Cauble, P. Celliers, G. Collins, L. Da Silva, T. Hall, M. Moret, M. Sacchi, P. Baclet, B. Cathala, Equation of state data for iron at pressures beyond 10 Mbar. Phys. Rev. Lett. **88**, 235502 (2002)
78. A.L. Kritcher, T. Döppner, D. Swift, J. Hawreliak, G. Collins, J. Nilsen, B. Bachmann, E. Dewald, D. Strozzi, S. Felker, O.L. Landen, O. Jones, C. Thomas, J. Hammer, C. Keane, H.J. Lee, S.H. Glenzer, S. Rothman, D. Chapman, D. Kraus, P. Neumayer, R.W. Falcone, Probing matter at Gbar pressures at the NIF. High Energy Density Phys. **10**, 27–34 (2014)
79. M.A. Purvis, V.N. Shlyaptsev, R. Hollinger, C. Bargsten, A. Pukhov, A. Prieto, Y. Wang, B.M. Luther, L. Yin, S. Wang, J.J. Rocca, Relativistic plasma nanophotonics for ultrahigh energy density physics. Nat. Photonics **7**, 796–800 (2013)
80. A. Di Piazza, C. Müller, K.Z. Hatsagortsyan, C.H. Keitel, Extremely high-intensity laser interactions with fundamental quantum systems. Rev. Mod. Phys. **84**, 1177–1228 (2012)
81. Surface plasmon resurrection (editorial). Nat. Photonics **6**, 707 (2012)
82. W.L. Barnes, A. Dereux, T.W. Ebbesen, Surface plasmon subwavelength optics. Nature **424**(6950), 824–830 (2003)
83. D.K. Gramotnev, S.I. Bozhevolnyi, Plasmonics beyond the diffraction limit. Nat. Photonics **4**, 83–91 (2010)
84. A.G. Brolo, Plasmonics for future biosensors. Nat. Photonics **6**(11), 709–713 (2012)
85. E. Ozbay, Plasmonics: merging photonics and electronics at nanoscale dimensions. Science **311**(5758), 189–193 (2006)
86. V.A. Vshivkov, N.M. Naumova, F. Pegoraro, S.V. Bulanov, Nonlinear electrodynamics of the interaction of ultra-intense laser pulses with a thin foil. Phys. Plasmas **5**(7), 2727–2741 (1998)
87. J.M. Pitarke, V.M. Silkin, E.V. Chulkov, P.M. Echenique, Theory of surface plasmons and surface-plasmon polaritons. Rep. Prog. Phys. **70**(1), 1 (2007)
88. S.A. Maier, *Plasmonics: Fundamentals and Applications* (Springer, New York, 2007)
89. A. Sgattoni, L. Fedeli, G. Cantono, T. Ceccotti, A. Macchi, High field plasmonics and laser-plasma acceleration in solid targets. Plasma Phys. Control. Fusion **58**(1), 014004 (2016)

90. A. Ono, H. Sano, W. Inami, Y. Kawata, Surface plasmon excitation and localization by metal-coated axicon prism. Micromachines **3**(1), 55 (2012)
91. N. Vasilantonakis, M.E. Nasir, W. Dickson, G.A. Wurtz, A.V. Zayats, Bulk plasmon-polaritons in hyperbolic nanorod metamaterial waveguides. Laser Photonics Rev. **9**(3), 345–353 (2015)
92. H.H. Nguyen, J. Park, S. Kang, M. Kim, Surface plasmon resonance: a versatile technique for biosensor applications. Sensors **15**(5), 10481–10510 (2015)
93. D.K. Gramotnev, S.I. Bozhevolnyi, Nanofocusing of electromagnetic radiation. Nat. Photonics **8**, 13–22 (2014)
94. D.W. Pohl, UCh. Fischer, U.T. Dürig, Scanning near-field optical microscopy (SNOM). J. Microsc. **152**(3), 853–861 (1988)
95. C.M. Soukoulis, M. Wegener, Past achievements and future challenges in the development of three-dimensional photonic metamaterials. Nat. Photonics **5**, 523–530 (2011)
96. A.V. Kildishev, A. Boltasseva, V.M. Shalaev, Planar photonics with metasurfaces. Science **339**(6125) (2013)
97. B. Liedberg, C. Nylander, I. Lunström, Surface plasmon resonance for gas detection and biosensing. Sens. Actuators **4**, 299–304 (1983)
98. S. Pillai, K.R. Catchpole, T. Trupke, M.A. Green, Surface plasmon enhanced silicon solar cells. J. Appl. Phys. **101**(9), 093105 (2007)
99. T. Ceccotti, V. Floquet, A. Sgattoni, A. Bigongiari, O. Klimo, M. Raynaud, C. Riconda, A. Heron, F. Baffigi, L. Labate, L.A. Gizzi, L. Vassura, J. Fuchs, M. Passoni, M. Květon, F. Novotny, M. Possolt, J. Prokůpek, J. Proška, J. Pšikal, L. Štolcová, A. Velyhan, M. Bougeard, P. D'Oliveira, O. Tcherbakoff, F. Réau, P. Martin, A. Macchi, Evidence of resonant surface-wave excitation in the relativistic regime through measurements of proton acceleration from grating targets. Phys. Rev. Lett. **111**, 185001 (2013)
100. L. Fedeli, A. Sgattoni, G. Cantono, I. Prencipe, M. Passoni, O. Klimo, J. Proska, A. Macchi, T. Ceccotti, Enhanced electron acceleration via ultra-intense laser interaction with structured targets. Proc. SPIE **9514**, 95140R–95140R-8 (2015)
101. L. Fedeli, A. Sgattoni, G. Cantono, D. Garzella, F. Réau, I. Prencipe, M. Passoni, M. Raynaud, M. Kveton, J. Proska, A. Macchi, T. Ceccotti. Electron acceleration by relativistic surface plasmons in laser-grating interaction. Phys. Rev. Lett. **116**, 015001 (2016)
102. B. Eliasson, Instability of a thin conducting foil accelerated by a finite wavelength intense laser. New J. Phys. **17**(3), 033026 (2015)

Chapter 3
Numerical Tools

As in many research fields, numerical simulations play an important role in plasma physics. Indeed, plasmas are complex physical systems and in several scenarios analytical theories are of limited applicability. Numerical simulations are often required to clarify the physical processes at play in certain conditions or to prepare experimental activities.

This chapter is mainly devoted to present Particle-In-Cell (PIC) codes. This simulation technique is widely used in plasma physics to study a great variety of topics, ranging from astrophysical scenarios to laser-plasma interaction.

Section 3.1 gives a brief overview of numerical simulations in plasma physics, whereas Sect. 3.2 deals more specifically with PIC codes (although most of the fundamental algorithms are described in detail in Appendix B). The open-source PIC code *piccante*, mainly developed at University of Pisa, is presented in detail in Sect. 3.3. Another PIC code, *piccolino*, based on a different Maxwell solver, is presented in Sect. 3.4. A few applications of these codes are briefly discussed in Sect. 3.5.

The research activity presented in the next chapters is heavily based on numerical simulations performed with *piccante* code.

3.1 Numerical Simulations of Plasma Physics

Numerical simulations are an invaluable tool in the exploration of complex, nonlinear plasma phenomena. A wide range of numerical methods has been developed to simulate plasma in a variety of physical scenarios, from diluted space plasmas [1] to warm dense matter [2]. Depending on the regime, different numerical tools are appropriate for the description of the physical processes at play. Figure 3.1 (adapted from [3]) provides an overview of plasma models and related simulation tools, ranging from a fine-grained, microscopic description to a coarse-grained fluid description. A thorough review of the available simulation tools in plasma physics is however beyond the scope of this document.

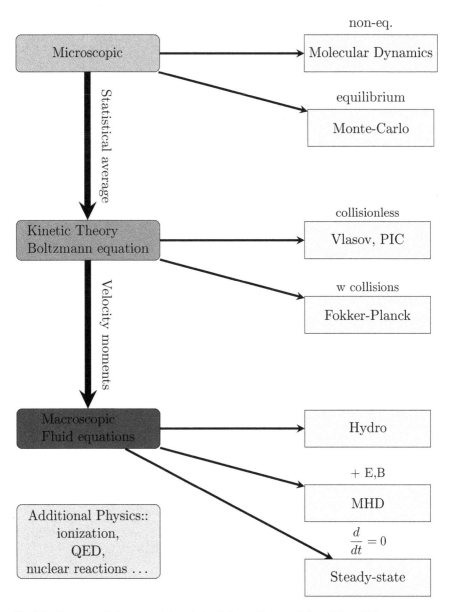

Fig. 3.1 Overview of plasma models and simulation techniques (adapted from [3])

This thesis deals essentially with very high-intensity laser plasma interaction on a time scale of few hundreds of femtoseconds at most. In these conditions, a kinetic description of the plasma is usually needed. Since in several cases collisional effects play a negligible role, the relativistic Vlasov equation coupled with Maxwell's equations is an appropriate theoretical description of these physical processes (see Sect. 2.2). Two main numerical tools are used to model these scenarios: *Vlasov codes* [4], which directly solve the Vlasov–Maxwell system of equations, and *Particle-In-Cell codes* [5, 6], in which the distribution functions of the particle species are sampled with a collection of macro-particles.

Compared to PIC codes, Vlasov codes usually perform better in resolving small features of the particle distribution functions (see for instance [7, 8]). However, multi-dimensional Vlasov simulations require a very large amount of computational resources, which can easily exceed the limits of state-of-the-art High Performance Computing facilities for fully 3D simulations.[1]

Vlasov codes won't be covered further in the present manuscript. Instead, since most of the numerical work presented in this thesis relies on simulations of laser-matter interaction performed with PIC codes, their working principle will be described in the next section. The interested reader can find a comprehensive reference on PIC algorithms in [5] and a recent review, which highlights modern developments, in [10].

3.2 Particle-In-Cell Codes

A PIC code solves the relativistic Vlasov equation (see Sect. 2.2) for each particle species:

$$\partial_t f_i + \mathbf{v} \cdot \nabla_x f_i + q \left(\mathbf{E} + \frac{1}{c} \mathbf{v} \times \mathbf{B} \right) \cdot \nabla_p f_i = 0 \tag{3.1}$$

together with Maxwell's equations for the EM field:

$$\begin{cases} \partial_t \mathbf{B} = -c \nabla \times \mathbf{E} & (3.2) \\ \partial_t \mathbf{E} = c \nabla \times \mathbf{B} - 4\pi \mathbf{J} & (3.3) \\ \nabla \cdot \mathbf{E} = 4\pi \rho & (3.4) \\ \nabla \cdot \mathbf{B} = 0 & (3.5) \end{cases}$$

The main idea behind Particle-In-Cell simulations is to sample the plasma distribution function with several charged macroparticles, while the electromagnetic field and the current are calculated on a grid.

The values of the fields acting on each macroparticle are interpolated from the field values at the grid nodes. Positions and momenta of the macroparticles are advanced in time using these field values at each time step. A current density at each

---

[1]Suppose that each one of the six dimensions x, y, z, p_x, p_y, p_z is resolved with 10^3 grid points. Supposing that each grid node requires 4 bytes of memory (a standard single-precision *float* number) to be represented, the total memory requirement amounts to $\sim 4 \cdot 10^6$ Terabytes! This is approximately 4000 times the total RAM available at the top # 1 supercomputer in the world [9], as of June 2015.

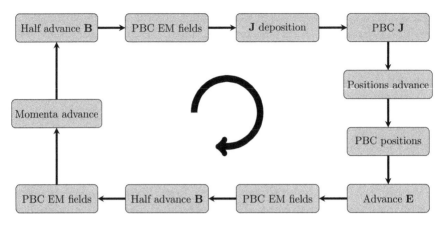

Fig. 3.2 Scheme of Particle-In-Cell algorithm

node is calculated from the moving macroparticles. A Maxwell solver advances in time the EM field equations, using the current density deposed on the grid by the macroparticles. Figure 3.2 illustrates synthetically the structure of a PIC simulation time-step.

Typically, the so-called *Boris pusher* is used to advance particle positions in time. Two different strategies are commonly implemented to calculate current deposition on the grid: an energy-conserving algorithm and a charge-conserving algorithm (the so-called *Esirkepov* current deposition). The first strategy enforces energy conservation at the expense of a non-exact conservation of the total charge (i.e. a net charge may be deposed on the grid, leading to possible non-physical effects). The Esirkepov current deposition strategy instead ensures exact conservation of the total charge, though at the expense of energy conservation. Charge-conserving algorithm is significantly more expensive than energy-conserving algorithm in terms of computational resources. In a typical simulation the differences between those strategies should be limited, however some scenarios may require specifically one of these strategies to be simulated correctly (see [11] for energy and momentum conservation in PIC codes). Finally, a typical Maxwell solver used in PIC simulations is the second order FDTD solver on a yee-lattice. Other higher-order FDTD Maxwell solvers or solvers based of Fast Fourier Transform may be used in some situations. In Appendix B the aforementioned algorithms are discussed in more detail.

If the standard yee-lattice FDTD scheme is used for the Maxwell solver, the so-called *Courant condition* should be satisfied in order to ensure numerical stability:

$$\Delta t < \frac{1}{c\sqrt{\frac{1}{(\Delta x)^2} + \frac{1}{(\Delta y)^2} + \frac{1}{(\Delta z)^2}}} \quad (3.6)$$

where Δt is the time-step size and Δx, Δy and Δz are the grid spacing.

3.2 Particle-In-Cell Codes

The core numerical techniques for PIC codes are well established, although there's an ongoing effort to add more physical processes to the model or to introduce new features. For instance, PIC codes including a physical model for ionization are relatively common [12]. Several PIC codes implement also particle collisions [13, 14], which are important in some physical scenarios. Taking into account radiative losses when ultra-relativistic particles are involved can also be important, since the back-reaction on the particles may greatly affect its dynamics (see [15]). This is especially important to simulate extremely relativistic astrophysical scenarios or laser-matter interaction in the intensity regime which will be available at the next-generation laser facility. Finally, the inclusion of QED effects (see [16]) is currently a very active research topic.

Supercomputing facilities now available allow to perform very large 3D simulations involving several 10^{10} macroparticles. PIC numerical scheme is inherently suitable for large scale parallel implementation: the simulation area can be split in several smaller region, each one of these is associated with a different process. Only information concerning EM fields and current at the borders and on particles crossing the borders should be exchanged between the processes. Some specific tasks however become increasingly demanding in massively parallel simulations (e.g. output routines).

PIC codes are at the forefront of High Performance Computing and there is ongoing effort to develop efficient PIC codes for new HPC architectures, such as those based on Graphical Processing Units (see [17, 18]).

3.3 PICCANTE: An Open-Source PIC Code

Piccante is an open-source, massively parallel, fully relativistic particle-in-cell code. *piccante* has been mainly developed by L.Fedeli(University of Pisa and CNR-INO) and A.Sgattoni(CNR-INO),[2] with significant contributions by S.Sinigardi(University of Bologna). Recently the project has benefited from the support of a dedicated Preparatory PRACE (PaRtnership for Advanced Computing in Europe) project, aimed at improving the efficiency of some core routines (see [19]).

piccante has been developed essentially from scratch, aiming at flexibility (in order to perform a wide variety of numerical simulations) and scalability (in order to perform large scale 3D PIC simulations on supercomputing facilities). The code is entirely written in C++. *piccante* was designed to run on a wide range of computing machines, from laptops to high performance supercomputing machines. The code was tested on up to 32768 parallel processes and it has proven to be well scalable [19]).

The code was cross-checked with another PIC code available in the group (*ALaDyn*, developed at the University of Bologna [20]) and the capability to reproduce known analytical results (wakefield excitation in laser interaction with underdense plasma, growth rate of instabilities, 1D solitons ...) was tested.

---

[2]Also affiliated at Politecnico di Milano when the project started.

Fig. 3.3 *piccante* logo. The logo is distributed under a Creative Commons attribution license http://creativecommons.org/licenses/by/4.0/

Although a lot of Particle-In-Cell codes have been developed by several groups in the world, only a small percentage is available to the whole community and only a fraction of that percentage is available with a proper GPLv3 open source license. *piccante* was released to the whole community as an open-source project (see [21]). Making available the code used for scientific simulations should ensure the reproducibility of the data (see [22]), which should be a cardinal tenet in modern science (Fig. 3.3).

The key feature of *piccante* code are reported below (see also Fig. 3.4, where some of these features are represented graphically).

piccante allows simulations with 1D, 2D and 3D Cartesian geometries. The simulation grid should be rectangular (i.e. Δx, Δy and Δz may be different)[3] Moreover, grid stretching along $\hat{x}$, $\hat{y}$ and $\hat{z}$ is possible. This allows the simulation grid to be finely resolved at its central region and less resolved at the edges of the box. This strategy saves computational resources (essentially memory) while keeping high the resolution where it is required. However, grid stretching leads to a significant increase of computational time for current deposition and field interpolation. Thus the opportunity of using this feature should be considered carefully.

As far as boundary conditions are of concern, it is possible to choose between Periodic Boundary Conditions (PBC) and Open Boundary Conditions (OBC).

Parallelization is done slicing the simulation box. It is possible to parallelize the simulation along $\hat{x}$, $\hat{y}$ and $\hat{z}$.

Finally, *piccante* allows to perform simulations with a moving window, which is useful in several cases (a paradigmatic case is a Laser Wakefield simulation).

piccante is a fully relativistic electromagnetic code.

---

[3]This is often useful in order to save computational time, since in typical simulations longitudinal resolution is more critical than transversal resolution. In this case Δx can be smaller than Δy and Δz.

3.3 PICCANTE: An Open-Source PIC Code

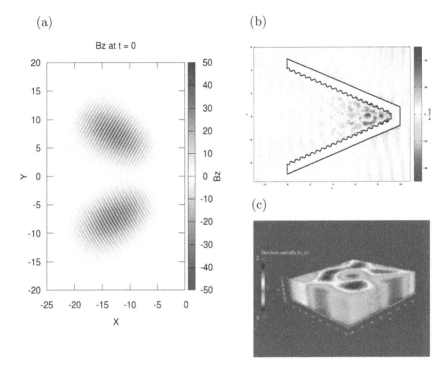

Fig. 3.4 Some of the features of *piccante*. **a** Multiple laser pulses. **b** Complex targets. **c** 3D simulations

The code allows to enable Radiation Reaction effects, which essentially take into account radiative losses typical of extremely relativistic scenarios. Radiation Reaction losses (see Sect. 2.2.6) are implemented in *piccante* following the algorithm described in [15].

It is possible to choose between several temperature distributions (Waterbag, Supergaussian and Maxwell) for the particle species. Moreover, it is possible to boost a temperature distribution. If a plasma is moving with a bulk velocity **v** and if, for instance, a Maxwellian temperature distribution is required, the momentum distribution is Maxwellian in a reference frame co-moving with the plasma. This last feature is particularly useful for astrophysical scenarios, where a temperature distribution may be assigned to a group of particles moving with a relativistic bulk velocity.

piccante has been designed to be very flexible in order to simulate a wide range of physical scenarios.

Complex target geometries can be easily implemented. A set of frequently used targets is provided (i.e. uniform boxes, targets with a soft ramp ...) but, as long as the target density can be written as a function $\rho(x, y, z)$, the implementation of new target geometries is straightforward. Chapter 5 provides an example of a non-standard initial geometry for a plasma. In this case the code was modified to load a pre-calculated external file, containing the coordinates and radii of a collection of spheres. These spheres were arranged in space in a way which mimicked the structure of a nanostructured foam targets.

An unlimited number of laser pulses can be placed in the simulation box and an unlimited number of particle species can be used (a particle species should be of one of these types: electrons, positrons or ions with selectable Z/A ratio).

piccante implements an advanced management of its output. Output timing can be selected in a flexible and straightforward way. Moreover the output can be easily restricted to a subregion of the simulation box (which is extremely useful when only a small region of a large simulation is of interest). Two output classes exist: lightweight output (containing synthetic information on the simulation in human-readable format) an heavyweight output (containing charge distributions, electromagnetic fields …in binary format). The typical size of a lightweight output file is a few kilobytes, while the size of an heavyweight output is usually in the hundreds of megabytes-few gigabytes range for 2D simulations and in the tens or hundreds of gigabytes for 3D simulations.

Since the binary output of the code is not in a standard format, a complete set of tools for the analysis of the simulation results is provided [23]. These tools allow to extract information from *piccante* output in a file format suitable for scientific visualisation software (*Paraview* [24], *Visit* [25], *Gnuplot* [26]).

Output was carefully optimized to achieve good scalability up to tens of thousands of MPI tasks [19] (see Sect. 3.3.1 for more details on this topic).

For typical simulations, the code can be controlled entirely from an input file written in *JSON* (JavaScript Object Notation) format. Only if custom plasma shapes are needed the user is required to manipulate the source code and recompile the program.

3.3.1 Optimization of Piccante

piccante code underwent an extensive rewriting of its output routines within the framework of a *Preparatory Prace* project. Details of the performed optimizations are reported in [19]. All the optimization work was performed on FERMI IBM BlueGene/Q machine, hosted at CINECA, Italy (Fig. 3.6 shows the modular architecture of the HPC machine, from the 16 core chip up to the whole 2 PFlops/s system). However, since the optimizations are not specifically tailored for the BG/Q architecture, a performance increase is expected also on similar parallel architectures (some preliminary tests on superMUC machine, based on intel CPUs, showed increased performances for the optimized *piccante* code). Most of the optimization work was targeted on output strategies, since profiling performed with *scalasca* revealed that for more than a few thousands MPI-tasks output routines became a severe bottleneck. Figure 3.5 shows the results of a profiling performed before and after optimization

3.3 PICCANTE: An Open-Source PIC Code

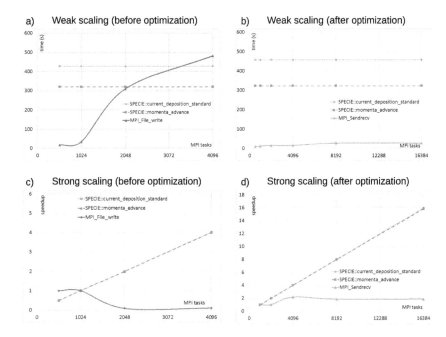

Fig. 3.5 Weak and strong scaling of *piccante* code on JUQUEEN supercomputers. Picture reproduced from [19]

(both weak scaling and strong scaling results are shown).[4] It is evident from the graphs that before optimization the output routines required a sizeable fraction of the code execution time.

The old (pre-optimization) and the new (post-optimization) output strategies are schematically illustrated in Fig. 3.7. The old output strategy relied upon the `MPI_File_write` routine (i.e. the routine for parallel output built in the standard MPI library implementation). With this strategy, for every output request, all the MPI tasks write "simultaneously" in the same shared file. This method proved to be efficient for a relatively small number of MPI tasks ($\lesssim 1024$). However, for larger numbers of MPI tasks `MPI_File_write` routines became a severe bottleneck. Indeed, for a BlueGene/Q machine, for every 128 compute notes there is only one I/O node communicating directly to the parallel I/O subsystem. This means that a single CPU deputed to I/O operations (16 CPU per I/O node) has to deal with the concurrent requests of 128 MPI tasks (even up to 512 if simultaneous multi-threading is

---

[4]These tests were performed with a 3D box completely filled with a uniform, low-temperature plasma. The simulation box was evenly split between the MPI tasks. In a weak scaling test the computational cost per MPI task is kept constant (i.e. if the number of MPI tasks is increased $2\times$, the simulation box is enlarged $2\times$).Therefore, in the ideal case, the execution time should be the same for any number of MPI tasks. In the strong scaling tests instead, the size of the simulation box is kept constant. Thus increasing $N\times$ the number of MPI tasks should lead to a $N\times$ speed-up in the ideal case.

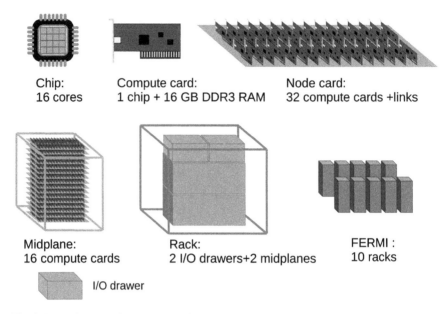

Fig. 3.6 Architecture of the BlueGene/Q machine FERMI

used). This large number of concurrent requests results in a very inefficient utilization of the output bandwidth.

Several parallel output strategies were developed and tested. A first significant improvement can be achieved using a different built-in function for parallel output, MPI_File_write_all instead of MPI_File_write. MPI_File_write_all is a collective output routine (all the MPI tasks involved should call the routine simultaneously) and thus additional optimizations are enabled. This solution improved the writing times significantly, but still maintained a very bad scaling for numbers of MPI tasks greater than 2048.

Another strategy we tested was grouping the MPI tasks into "writing groups" of various sizes. Within each group, only one MPI task was responsible for writing all the the data of its group. All the writing tasks concurrently write their data in a single large file. Despite actually improving performances with respect to naive approach based on MPI_File_write, even this scheme failed to show good scalability (as the number of MPI tasks grows, the number of concurrent writing tasks grows as well).

We tested also another naive approach, base on a single output file per MPI task. This strategy proved to scale well, though not being the fastest strategy for $N < 2048$. However, the very large number of output files is a severe limitation for post-processing.[5]

---

[5]Even listing the content of a directory containing tens of thousands of files may take several seconds or tens of seconds on FERMI.

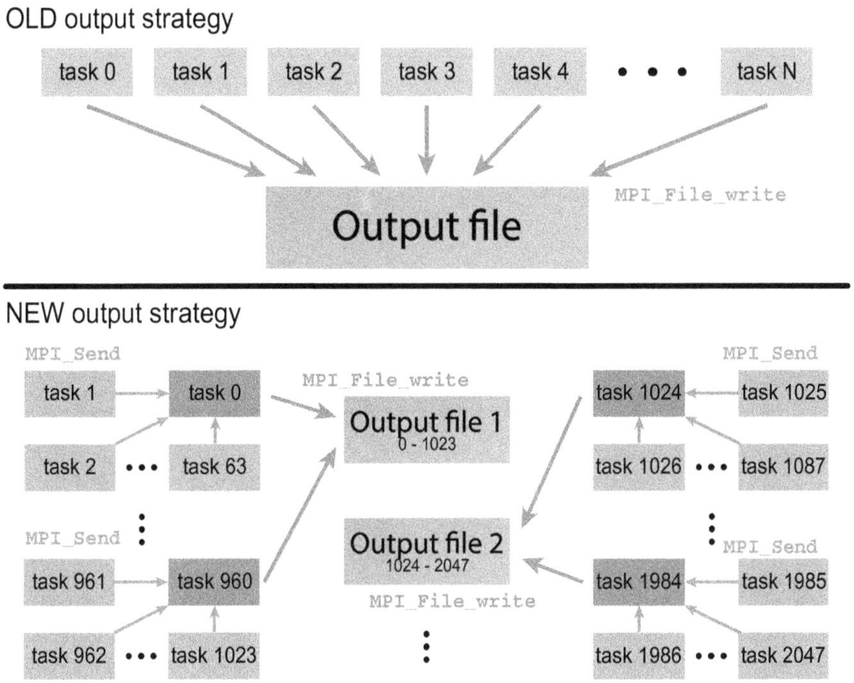

3.3 PICCANTE: An Open-Source PIC Code

OLD output strategy

NEW output strategy

Fig. 3.7 Scheme of the old and new output strategies in *piccante* is shown. In this example, G = 64 and F = N/128. Picture reproduced from [19]

The final solution which proved to be both efficient and scalable is illustrated in Fig. 3.7, in the lower panel. Let N be the total number of MPI tasks and G the number of writing groups. Within each group MPI tasks send their data to a master task. Only the M (M = N/G) master tasks are responsible for direct I/O operations, using the `MPI_File_write` routine. The output data is split into F different files (F < M) and M/F master tasks share a single output file. We tested several values for G ranging from G = 32 up to G = 128. For F we tested F = N/1024 and F = N/2048 cases. The new I/O strategy led to a major speed-up of writing operations. We observed up to 40× speed-up for particle data (the phase-space output) and up to 600× for field data (the EM field or the current or the charge density). Figure 3.8 provides a detailed comparison of the output time for the old and the new output strategy. The graph shows that not only the output times are orders of magnitude lower, but also the scaling with the number of MPI tasks is significantly improved. With the new output strategy we were able to reach an output speed equal to 8 GB/s, which is ∼50 % of the average output bandwidth.

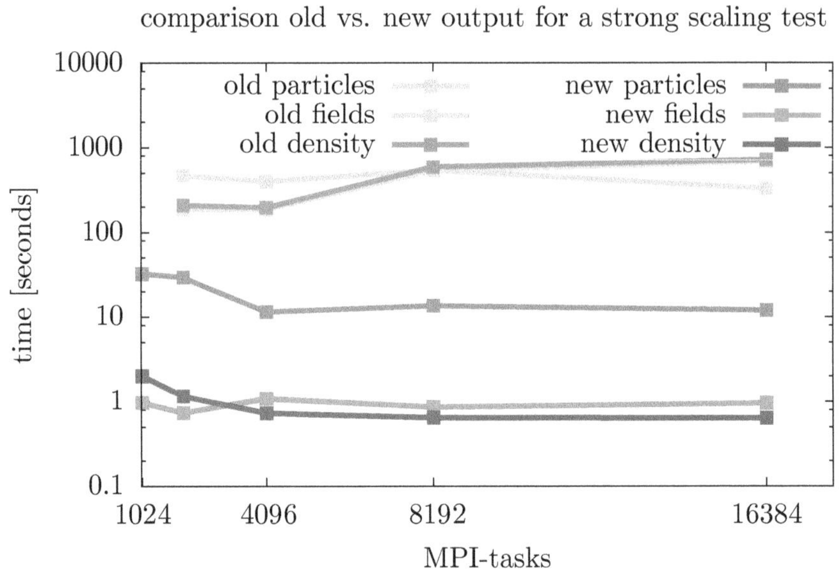

Fig. 3.8 Comparison of the output times between the old (*empty symbols* in light shades of *red* and *blue*) and the new strategies (*filled symbols* in *dark shades* of *red* and *blue*) versus the number of MPI-tasks N. The time is reported in seconds and corresponds to the aggregated time given by *scalasca* profiler dived by the number of MPI tasks N. The time spent to write particle positions and momenta on disk is represented in *red*, while the total time spent to write EM fields and charge densities is represented in *blue*. Picture reproduced from [19]

3.4 PICcolino: A Spectral PIC Code

piccante, as most of the other existing PIC codes, adopts a second order FDTD yee-lattice method to solve Maxwell equations on the computational grid. Though appropriate for a wide range of plasma phenomena and very efficient in terms of execution time, this scheme has a few severe weaknesses. For instance, if the numerical dispersion relation is calculated for EM waves propagating in the vacuum, a phase velocity slightly smaller than the speed of light is found (the exact phase velocity depends on the grid resolution, the time-step size and the propagation direction). This may cause unwanted physical effects in physical scenarios in which reproducing faithfully the phase of a laser pulse propagating in a plasma is crucial (i.e. Laser Wakefield simulations). Moreover, in some scenarios involving ultra-relativistic particles, the velocity of a macro-particle can actually be greater than the numerical phase velocity of light. This leads to a non-physical numerical Cherenkov effect; even a macro-particle travelling at constant speed in the vacuum can couple to the EM field and emit radiation. The numerical Cherenkov effect can severely affect the particle dynamics, making the simulation meaningless.

3.4 PICcolino: A Spectral PIC Code

The calculated dispersion relation for EM waves in vacuum in 3D is (see [27]):

$$\left[\frac{1}{c\Delta t}\sin\left(\frac{\omega\Delta t}{2}\right)\right]^2 = \left[\frac{1}{\Delta x}\sin\left(\frac{k_x\Delta x}{2}\right)\right]^2 + \left[\frac{1}{\Delta y}\sin\left(\frac{k_y\Delta y}{2}\right)\right]^2 + \left[\frac{1}{\Delta z}\sin\left(\frac{k_z\Delta z}{2}\right)\right]^2 \tag{3.7}$$

where c is the speed of light, k is the wavenumber and ω is the frequency, Δ_x, Δ_y, Δ_z are the grid spacings in the three dimensions and Δt is the time-step. For simplicity we consider a 2D simulation (the third term of the right hand side of Eq. 3.7) with the same resolution on $\hat{x}$ and $\hat{y}$, thus $\Delta x = \Delta y$. Using a Courant factor $f_c = 0.98$, from Eq. 3.6 we get $\Delta t = \frac{f_c}{\sqrt{2}}\Delta x$. Finally, since $c = 1$, Eq. 3.7 becomes:

$$\sin^2\left(\frac{\omega f_c \Delta x}{\sqrt{2}}\right) = \frac{f_c^2}{2}\left[\sin^2\left(\frac{k_x\Delta x}{2}\right) + \sin^2\left(\frac{k_y\Delta x}{2}\right)\right] \tag{3.8}$$

Using the previous results, the phase velocity $v_f = \frac{\omega}{k}$ can be written as:

$$v_f(k_x, k_y) = \frac{2\sqrt{2}}{f_c\Delta x} \frac{\arcsin\left(\sqrt{\sin^2\left(\frac{k_x\Delta x}{2}\right) + \sin^2\left(\frac{k_y\Delta x}{2}\right)}\right)}{\sqrt{k_x^2 + k_y^2}} \tag{3.9}$$

Figure 3.9 shows $v_f(k_x, k_y)$ for $\Delta x = 0.05$. It is evident from the graph that for larger k the error on the phase velocity increases and that the issue is more serious on the axes. For $\mathbf{k} = 2\pi(1, 0)$ the error is ∼0.35 %. Thus, an ultra-relativistic particle with $\gamma > 143$ will have a velocity $v > v_f$.

Several techniques exist to keep this unwanted behaviour under control. A few techniques are based on higher order algorithms for the Maxwell solver. Another strategy is based on the use of a spectral solver. Transforming the EM field in its Fourier components, evolving in time these components and transforming back to the real space, allow the EM waves to propagate at the speed of light, virtually eliminating the numerical Cherenkov effect. This topic is discussed in detail in [28].

piccolino was initially developed mainly as an numerical exercise, but it has evolved into a valid alternative to other PIC codes available in the group for numerical simulations involving ultra-relativistic particles. In order to perform Fast Fourier Transform efficiently, *FFTW (the Fastest Fourier Transform of the West)* library was used [29]. Parallelization of the most computationally expensive routines (current deposition, particle pusher) is performed with OpenMP library (i.e. the operations are carried out by different threads sharing the whole memory). Due to the adopted parallelization technique, *piccolino* is not suitable for large supercomputers, being mainly designed to run on personal computers or workstations.

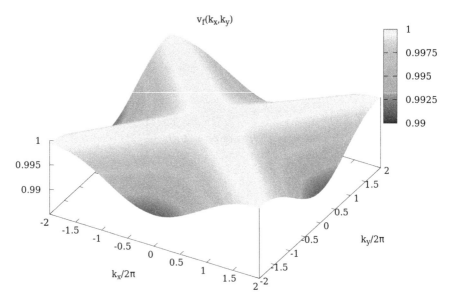

Fig. 3.9 The figure shows the phase velocity v_f of an EM wave in a simple FDTD solver as a function of $k_x/(2\pi)$ and $k_y/(2\pi)$. The graph is calculated for $\Delta x = 0.05\lambda$. It is evident from the graph that the largest error in the phase velocity is for the propagation along the axes

3.5 Applications

piccante was extensively used to support the research activity presented in this dissertation. This section is intended to present concisely other applications of *piccante* and *piccolino* code, not related to High Field Plasmonics, in order to show the flexibility of these computational tools. Section 3.5.1 provides an example in which the investigation of an astrophysical scenario (Weibel instability in counter-streaming relativistic pair-plasmas) required the use of both codes. Section 3.5.2 concerns instead numerical simulations of ion acceleration with Coulomb explosion of thin gold foils (these simulation were performed using only *piccante* code).

3.5.1 Weibel Instability in Pair-Plasmas

In a recently published work (see [30]), the development of Weibel instability generated by two counter-streaming, ultra-relativistic, pair plasmas (electrons and positron) is studied. The topic is a well known scenario in plasma physics and astrophysics and has been explored extensively both from the analytical point of view and the numerical point of view (see for instance [31]). Our main contribution is to investigate the role of radiation reaction in these conditions.

3.5 Applications

The study of a model problem characterized by two colliding electron-positron plasma clouds at relativistic velocities is relevant for several astrophysical scenarios, including the fireball model of Gamma Ray Bursts [32], pulsar wind outflows in Pulsar Wind Nebulae [33], and relativistic jets from Active Galactic Nuclei [32].

3.5.1.1 Numerical Simulations with PICCANTE and PICcolino

The initial configuration of the 2D numerical simulations consists in two neutral beams of electron-positron pairs propagating in opposite directions and filling the whole simulation space (as in the first panel of Fig. 3.10). The initial γ factor of the particles is 200 (with a low-temperature Maxwellian distribution, in order to provide initial noise). Two configurations were tested: the P-mode configuration, in which the initial velocity of the beams is along the simulation box, and T-mode configuration, in which the initial velocity is perpendicular to the simulation box. These configurations are prone to a host of instabilities, characterized by the orientation of their wave-vector with respect to the direction of the beams (see [34]).

Two limiting cases for the instability exist: the longitudinal two stream instability (TSI), corresponding to an electrostatic mode with a wave-vector aligned with the particle flow, and the transverse filamentation instability (FI), corresponding to an EM mode with wave-vector perpendicular to the beam direction. In the general case, the wave-vector is oblique with respect to the beam direction and the instability is a "mixture" of TSI and FI. However, analytical calculations based on first-order perturbation theory [34, 35], show that the growth rate Γ in the linear phase for the two-stream instability $\Gamma \propto \gamma_0^{-3/2}$, while for the FI $\Gamma \propto \gamma_0^{-1/2}$, where $\gamma_0 = \left(1 + (p_0/m_e c)^2\right)^{1/2}$ is the initial beam Lorentz factor (with p_0 the initial drift momentum). Moreover, these calculations show that, when the beams are symmetric, the instability is prevalently EM. Thus, in the ultra-relativistic regime the transverse FI is expected to dominate the growth of the instability, at least before nonlinear effects become important.

The instability leads to the formation of current filaments, which merge in larger structures during the nonlinear phase (see Fig. 3.10). After this coalescence phase, magnetic field and current filaments reach a quasi-stationary regime, with typical scales of several skin depths. Most particles are magnetically confined inside the current filaments. A group of particles are accelerated at twice their initial momentum, forming a peak in the energy spectrum. Both T-mode and P-mode simulations show a similar behaviour.

Numerical simulations for the T-mode case were performed with *piccante* code, using a simulation box 2000×2000 cells wide, with a resolution $\Delta x = \Delta y = 0.05\lambda_p$ (where λ_p is the plasma skin depth). For each species (two electron species and two positron species) 2×10^8 computational particles were used. The time-step size was $0.0325\,T_p$.

For the P-mode case, *piccante* could not be used, since its FDTD Maxwell solver is affected by severe numerical Cherenkov radiation when ultra-relativistic particles are involved. Thus for the P-mode simulations *piccolino* code was used (*piccolino* was benchmarked with *piccante* in test cases where numerical Cherenkov radiation

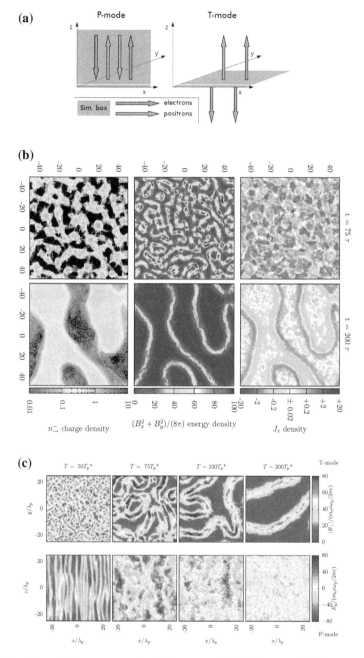

Fig. 3.10 **a** The two different geometries which were studied. **b** Charge density for a given species, energy density of the transversal magnetic field and longitudinal current for a P-mode simulation (two time-steps are shown). **c** The energy density of the transversal magnetic field for several time steps, comparing P-mode and T-mode simulations. **c** was adapted from [30]

3.5 Applications

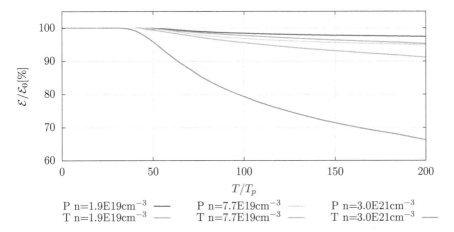

Fig. 3.11 Total energy as a function of simulation time for various plasma densities and simulation geometries (P and T modes). Radiation reaction is responsible for energy losses. The figure was reproduced from [30]

was negligible and the results were in full agreement). Simulations performed with *piccolino* were characterized by a smaller grid (1000×1000 cells, though the spatial resolution was identical) and a smaller time-step size ($0.025\, T_p$). The same number of particles per cell was used. Figure 3.11 shows the evolution in time of the total energy (particle energy plus field energy) normalized over the initial kinetic energy of the beams, for different simulations with RF included. Radiative losses are higher for T-mode (transversal simulations) than for the P-mode (longitudinal simulations), which is consistent with the higher fields generated in the T-mode case. Moreover, since the magnetic field saturation is proportional to the square of the total plasma density $\sqrt{n_T}$, higher densities correspond to larger radiative losses. RF losses account for a few percent of the initial energy for a plasma density of $10^{19}\,\text{cm}^{-3}$ and a simulation time of $100\, T_p$. For higher densities (not easily found in astrophysical scenarios involving pair plasmas), RF accounts for a major loss of energy. Despite this large energy loss, the dynamics of the instability is not strongly affected with respect to the case without RF. Even when RF is very important as far energy loss is of concern, we observe in the simulations the formation of filamentary structures hardly distinguishable from those obtained without RF. Two different factors contribute to preserving the main features of the instability: the EM fields have to grow in order for RF to be important (so RF plays little role before the saturation phase) and in the ultra-relativistic case the dominant term of RF force is $\propto \gamma^2$ (this means that RF affects mainly the particles in the high energy tail of the distribution).

3.5.2 Intense Laser Interaction with Thin Gold Targets

Recently, a promising ion acceleration scheme has been described in [36]. In this experimental work, ultrathin (nanometric) gold foils were irradiated at 8×10^{19} W/cm^2 (1.3 J on target) with a 30 fs Ti:Sapphire laser, obtaining high charge states (greater than Au^{50+}) and ion energies up to ∼1 MeV per nucleon. This result is remarkable because, as the authors claim, energies greater than 1 MeV/nucleon for heavy ions had been previously obtained only with significantly higher pulse energies on target (exceeding 20 J) [37–39]. The high efficiency of the ion acceleration process was attributed to a Coulomb explosion of the thin target.

Numerical simulations were performed with *piccante* code in order to support an experimental campaign carried out at VULCAN laser facility (Rutherford Appleton Laboratory, Didcot, Oxford, UK), aimed at testing the same acceleration scheme in a significantly more powerful laser facility. VULCAN is indeed a Petawatt-class laser, delivering 160–190 J on target in 700–900 fs and able to attain peak intensities of the order of 2–5×10^{20} W/cm^2 (the focal spot is close to the diffraction limit). Numerical simulations were performed in order to gain some insights on the acceleration mechanism in these conditions. Three large-scale 2D simulations were performed on the FERMI supercomputer. The main parameters of the numerical simulations are reported in Table 3.1. The differences between the simulations are highlighted (essentially the differences concern the details of the target composition and the laser pulse intensity). With respect to VULCAN laser parameters, the temporal length of the laser pulse was reduced to 400 fs in order to limit the computational cost of the simulations. For the same reason rather thick targets were simulated (50 nm instead of ∼15 nm used in [36]).[6]

A rather long laser-target interaction generally requires a large simulation box to avoid effects related to particles crossing the simulation borders (822 μm × 320 μm). On the other hand, a dense target requires a good spatial resolution to correctly reproduce the physical processes at play.[7] With these constraints, the stretching of the simulation grid is a particularly useful feature. The resolution is kept at the value quoted in Table 3.1 only in the interaction region (a 4 μm × 20 μm box), while the resolution is smoothly reduced towards the edges of the simulation box. Thanks to the stretched grid only 31002 × 8192 cells are used (instead of 98640 × 34880). Limited differences were observed between the three different simulated cases (Fig. 3.12).

The simulations suggest that the main physical process at play in these conditions is Coulomb explosion of the target. P-polarization indeed determines a strong coupling of the laser with the target: a significant amount of energy is transferred to the electrons and ion acceleration is due to their almost-symmetric expansion. The

---

[6]It is worth to stress that the main aim of these 2D simulations was to give an idea of the physical process at play, rather than reproducing experimental results faithfully.

[7]The skin depth should be resolved at least with one point. As a consequence at least 50 points per μm are required for 64 n_c densities. However with this resolution the 50 nm thin target would have been resolved with just 1–2 points. Thus a resolution of 120 points per μm in the longitudinal direction was chosen.

3.5 Applications

Table 3.1 Parameter list for 2D simulations of Vulcan experiment ($\lambda = 1\,\mu m$)

Simulation parameters	2D simulations
Simulation box	$822\,\mu m \times 320\,\mu m$
Resolution (interaction, points per μm)	[120, 109.6]
Target (bulk)	
Density	$60.0\,n_c$
Thickness	50 nm
Composition	**sim 1,2:** Au^{45+} (30 %) + $Au^{46+} - Au^{51+}$ (10 % each)
	sim 3: $Au^{43+} - Au^{52+}$ (10 % each)
Particles per cell (electrons)	100
Particles per cell (ions)	64 per species
Target (contaminants)	
Density	$6.0\,n_c$
Thickness	25 nm
Composition	C^{6+} (50 %) + H^+ (50 %)
Particles per cell (electrons)	100
Particles per cell (ions)	64 per species
Laser	
Polarization	P
a_0	**sim 1,3:** 12 (2×10^{20} W/cm^2)
	sim 2: 19 (5×10^{20} W/cm^2)
Waist	$4\,\lambda$
Type	Gaussian
Pulse incidence	$0°$
Duration FWHM	400 fs

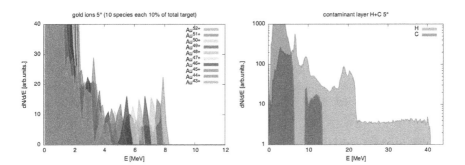

Fig. 3.12 Numerical ion spectra. The *left panel* shows gold ions while the *right panel* shows carbon and hydrogen ions. Data are shown for simulation 3 and for emission angle $5°$

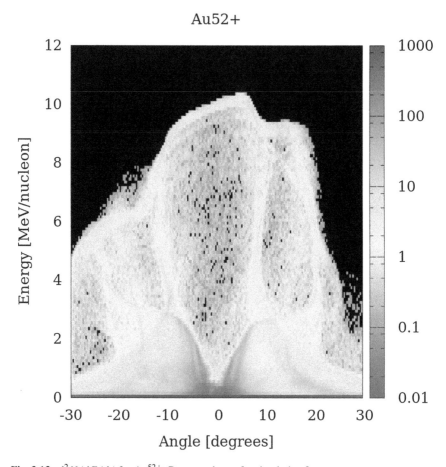

Fig. 3.13 $d^2N/dE/d\phi$ for Au^{52+}. Data are shown for simulation 2

energy spectra of the various ion species present a "bump-on-tail" (except for the protons). These narrow-band peaks are ordered according to their charge to mass ratio, consistently with a Coulomb explosion of the target (also in [36] an ordering of the heavy ion spectra according to their charge state was observed). The maximum ion energy is obtained off-axis (a typical energy-angle distribution for gold can be seen in Fig. 3.13).

References

1. J.F. Hawley, C.F. Gammie, S.A. Balbus, Local three-dimensional magnetohydrodynamic simulations of accretion disks. APJ **440**, 742 (1995)
2. B. Holst, R. Redmer, M.P. Desjarlais, Thermophysical properties of warm dense hydrogen using quantum molecular dynamics simulations. Phys. Rev. B **77**, 184201 (2008)

3. P. Gibbon, *Short Pulse Laser Interactions with Matter* (Imperial College Press, London, 2005)
4. N.J. Sircombe, T.D. Arber, Valis: a split-conservative scheme for the relativistic 2d Vlasov–Maxwell system. J. Comput. Phys. **228**(13), 4773–4788 (2009)
5. C.K. Birdsall, A.B. Langdon, *Plasma Physics via Computer Simulation* (CRC Press, Boca Raton, 2004)
6. J.M. Dawson, Particle simulation of plasmas. Rev. Mod. Phys. **55**, 403–447 (1983)
7. A. Grassi, L. Fedeli, A. Macchi, S.V. Bulanov, F. Pegoraro, Phase space dynamics after the breaking of a relativistic Langmuir wave in a thermal plasma. Eur. Phys. J. D, **68**(6), 1–8 (2014)
8. A. Grassi, L. Fedeli, A. Sgattoni, A. Macchi, Vlasov simulation of laser-driven shock acceleration and ion turbulence. Plasma Phys. Control. Fusion **58**(3), 034021 (2016)
9. top500, *Tianhe-2 (milkyway-2)* (National University of Defense Technology, 2015); (online). Accessed 22 Sept 2015
10. T.D. Arber, K. Bennett, C.S. Brady, A. Lawrence-Douglas, M.G. Ramsay, N.J. Sircombe, P. Gillies, R.G. Evans, H. Schmitz, A.R. Bell, C.P. Ridgers, Contemporary particle-in-cell approach to laser-plasma modelling. Plasma Phys. Control. Fusion **57**(11), 113001 (2015)
11. J.U. Brackbill, *On Energy and Momentum Conservation in Particle-in-Cell Simulation*. ArXiv e-prints (2015)
12. D.L. Bruhwiler, D.A. Dimitrov, J.R. Cary, E. Esarey, W. Leemans, R.E. Giacone, Particle-in-cell simulations of tunneling ionization effects in plasma-based accelerators. Phys. Plasmas **10**(5), 2022–2030 (2003)
13. C.K. Birdsall, Particle-in-cell charged-particle simulations, plus Monte Carlo collisions with neutral atoms, PIC-MCC. IEEE Trans. Plasma Sci. **19**(2), 65–85 (1991)
14. V. Vahedi, M. Surendra, A Monte Carlo collision model for the particle-in-cell method: applications to argon and oxygen discharges. Comput. Phys. Commun. **87**(1–2), 179–198 (1995); Particle Simulation Methods
15. M. Tamburini, F. Pegoraro, A. Di Piazza, C.H. Keitel, A. Macchi, Radiation reaction effects on radiation pressure acceleration. New J. Phys. **12**(12), 123005 (2010)
16. M. Lobet, E. d'Humières, M. Grech, C. Ruyer, X. Davoine, L. Gremillet, Modeling of radiative and quantum electrodynamics effects in PIC simulations of ultra-relativistic laser–plasma interaction. J. Phys. Conf. Ser. **688**(1), 012058 IOP Publishing (2016)
17. F. Rossi, P. Londrillo, A. Sgattoni, S. Sinigardi, G. Turchetti, Towards robust algorithms for current deposition and dynamic load-balancing in a gpu particle in cell code. AIP Conf. Proc. **1507**(1), 184–192 (2012)
18. H. Burau, R. Widera, W. Honig, G. Juckeland, A. Debus, T. Kluge, U. Schramm, T.E. Cowan, R. Sauerbrey, M. Bussmann, PIConGPU: a fully relativistic particle-in-cell code for a GPU cluster. IEEE Trans. Plasma Sci. **38**(10), 2831–2839 (2010)
19. A. Sgattoni, L. Fedeli, S. Sinigardi, A. Marocchino, A. Macchi, V. Weinberg, A. Karmakar, Optimising Piccante - An Open Source Particle-in-Cell Code for Advanced Simulations on Tier-0 Systems. Technical report, PRACE white papers (2015). (online). Accessed 23 May 2015
20. C. Benedetti, A. Sgattoni, G. Turchetti, P. Londrillo, ALaDyn: a high-accuracy pic code for the Maxwell–Vlasov equations. IEEE Trans. Plasma Sci. **36**(4), 1790–1798 (2008)
21. S. Sgattoni, L. Fedeli, S. Sinigardi, A. Marocchino, Piccante: a spicy massively parallel fully-relativistic electromagnetic 3D particle-in-cell code (2015). http://aladyn.github.io/piccante/
22. D.C. Ince, L. Hatton, J. Graham-Cumming, The case for open computer programs. Nature **482**(7386), 485–488 (2012)
23. S. Sgattoni, L. Fedeli, S. Sinigardi, A. Marocchino, Set of tools for reading piccante output files (2015). http://github.com/ALaDyn/tools-piccante
24. Kitware, Paraview 4.3: an open-source, multi-platform data analysis and visualization application (2015). http://www.paraview.org/
25. Lawrence Livermore National Laboratory, Visit 2.9.2: an open source, interactive, scalable, visualization, animation and analysis tool (2015). http://www.paraview.org/
26. T. Williams, C. Kelley, et al., Gnuplot 5.0: an interactive plotting program (2015). http://gnuplot.sourceforge.net/

27. W.H.A. Schilders, E.J.W.T.E.R. Maten, P.G. Ciarlet, *Numerical Methods in Electromagnetics: Special Volume. Handbook of Numerical Analysis* (Elsevier, Amsterdam, 2005)
28. P. Yu, X. Xu, V.K. Decyk, F. Fiuza, J. Vieira, F.S. Tsung, R.A. Fonseca, W. Lu, L.O. Silva, W.B. Mori, Elimination of the numerical Cerenkov instability for spectral em-pic codes. Comput. Phys. Commun. **192**, 32–47 (2015)
29. M. Frigo, S.G. Johnson, The design and implementation of FFTW3. Proc. IEEE **93**(2), 216–231 (2005); Special issue on "Program Generation, Optimization, and Platform Adaptation"
30. M. D'Angelo, L. Fedeli, A. Sgattoni, F. Pegoraro, A. Macchi, Particle acceleration and radiation friction effects in the filamentation instability of pair plasmas. Mon. Not. R. Astron. Soc. **451**(4), 3460–3467 (2015)
31. L.O. Silva, R.A. Fonseca, J.W. Tonge, J.M. Dawson, W.B. Mori, M.V. Medvedev, Interpenetrating plasma shells: near-equipartition magnetic field generation and nonthermal particle acceleration. Astrophys. J. Lett. **596**(1), L121 (2003)
32. M.C. Begelman, R.D. Blandford, M.J. Rees, Theory of extragalactic radio sources. Rev. Mod. Phys. **56**, 255–351 (1984)
33. P. Blasi, E. Amato, positrons from pulsar winds, in *High-Energy Emission from Pulsars and their Systems, Astrophysics and Space Science Proceedings*, ed. by Diego F. Torres, Nanda Rea (Springer, Berlin, 2011), pp. 623–641
34. A. Bret, L. Gremillet, M.E. Dieckmann, Multidimensional electron beam-plasma instabilities in the relativistic regime. Phys. Plasmas **17**(12), 120501 (2010)
35. F. Califano, F. Pegoraro, S.V. Bulanov, Spatial structure and time evolution of the Weibel instability in collisionless inhomogeneous plasmas. Phys. Rev. E **56**, 963–969 (1997)
36. J. Braenzel, A.A. Andreev, K. Platonov, M. Klingsporn, L. Ehrentraut, W. Sandner, M. Schnürer, Coulomb-driven energy boost of heavy ions for laser-plasma acceleration. Phys. Rev. Lett. **114**, 124801 (2015)
37. B.M. Hegelich, B. Albright, P. Audebert, A. Blazevic, E. Brambrink, J. Cobble, T. Cowan, J. Fuchs, J.C. Gauthier, C. Gautier, M. Geissel, D. Habs, R. Johnson, S. Karsch, A. Kemp, S. Letzring, M. Roth, U. Schramm, J. Schreiber, K.J. Witte, J.C. Fernández, Spectral properties of laser-accelerated mid-Z MeV/u ion beamsa). Phys. Plasmas **12**(5), 056314 (2005)
38. M. Hegelich, S. Karsch, G. Pretzler, D. Habs, K. Witte, W. Guenther, M. Allen, A. Blazevic, J. Fuchs, J.C. Gauthier, M. Geissel, P. Audebert, T. Cowan, M. Roth, Mev ion jets from short-pulse-laser interaction with thin foils. Phys. Rev. Lett. **89**, 085002 (2002)
39. B.M. Hegelich, B.J. Albright, J. Cobble, K. Flippo, S. Letzring, M. Paffett, H. Ruhl, J. Schreiber, R.K. Schulze, J.C. Fernández, Laser acceleration of quasi-monoenergetic MeV ion beams. Nature **439**(7075), 441–444 (2006)

Chapter 4
Electron Acceleration with Grating Targets

The main result described in this chapter is the first observation of electron acceleration by relativistic surface plasmons excited with ultra-high intensity laser pulses ($I > 10^{18}$ W/cm^2) interacting with grating targets. The experimental data are supported by three-dimensional numerical simulations and a theoretical model, which confirm the generation of relativistic surface plasmons and justify the acceleration process.

Two distinct experiments were carried out: an extensive campaign in 2014 with the UHI100 laser system (a 100 TW system located at CEA-Saclay, Paris, France) and a few measurements in 2015 on the PW-class laser system at GIST (Gwangju, Republic of Korea). Rather than constituting a complete experimental campaign on their own, the latter measurements were performed mainly to asses the feasibility of experiments with grating targets on a PW laser facility.[1]

The study of Surface Plasmons excitation and manipulation with structured materials is a vibrant research field, with a well understood theory (see Sect. 2.3). However, Surface Plasmons are normally excited at field intensities several orders of magnitude lower than $I = 10^{18}$ W/cm^2. In "traditional" plasmonics indeed, plasmonic structures are meant to withstand the interaction with the laser pulse, whereas in intense laser-matter interaction targets are completely destroyed during the interaction. The study of plasmonics effects in these conditions is essentially an unexplored ground: no general theory is known for this regime, in which relativistic, strongly nonlinear effects take place. Moreover, its experimental investigation has been hindered so far by pre-pulses inherent to high-power laser systems and causing early disruption of the target structures, which are required for surface plasmon excitation. In our experiments these difficulties are overcome using a system with an exceptionally "clean" laser pulse.

---

[1]These results obtained at GIST are still preliminary as of October 2015.

The electron acceleration mechanism described in this chapter provides ultrashort, brilliant electron bunches in the 10 MeV energy range. These characteristics are not easily attainable with other sources. Indeed, with conventional sources (e.g. electron guns) it is very difficult to obtain electron bunches shorter than $\sim$100 fs [1], while the temporal length of a laser-based source should be comparable to the duration of the pulse (few tens of fs). Of course with Laser Wake-Field Acceleration ultra-short electron bunches can be produced (see Sect. 2.2.7), but in this case obtaining final energies of only few MeVs with a good quality (peaked energy spectrum, high total charge) is challenging [2]. Besides being per se an attractive electron source at moderate energies (as will be detailed hereunder), the scheme we present is interesting because it could represent a first step to extend "traditional" plasmonics into the unexplored high field regime.

The experimental study of electron acceleration with laser-induced surface plasmons described in this chapter is, to our knowledge, the first one performed at fully relativistic laser intensities ($a_0 \simeq 5$).

In Sect. 4.1.1 an introduction setting the framework of the present work and describing some previous experimental and numerical activities with structured grating targets is given. Section 4.2 is devoted to a detailed exposition of the experimental activity performed at CEA-Saclay. In Sect. 4.3 the main results of the numerical simulation campaign are presented. The analytical model for electron acceleration with relativistic surface plasmons is then covered in Sect. 4.4. Section 4.5 is dedicated to the experimental data collected at GIST research center. Finally, Sect. 4.6 provides some conclusive remarks.

Some preliminary results of this experimental campaign can be found in [3], whereas the main results together with a thorough discussion can be found in [4].

4.1 Introduction and Previous Results

This section is provided in order to frame our work within the context of the ongoing research effort to study plasmonic effects with very high laser fields. Section 4.1.1 presents the main outcomes of a previous experimental investigation which paved the way for the research activity described in this chapter. Also some previous experiments performed with much lower laser intensities are briefly mentioned. Section 4.1.2 instead reviews some earlier numerical and theoretical investigations.

4.1.1 Previous Experimental Investigations

The research activity covered in this chapter was undertaken as a prosecution of a previous experimental investigation (carried out at the same laser facility), focused on enhanced TNSA with grating targets (see [5]). During this 2012 experimental campaign (carried out at CEA-Saclay), the interaction of intense fs laser pulses

4.1 Introduction and Previous Results

($I \sim 10^{19}$ W/cm^2) with thin grating targets was experimentally investigated. These targets were Mylar foils whose front surface was grooved with a grating spacing designed for an expected resonance at 30° pulse incidence. Also simple flat Mylar foils with approximately the same thickness were used for comparison.

Since the experiment was focused on ion acceleration, the main detector was a Thomson Parabola (the setup of the experiment is shown in Fig. 4.1). Together with the Thomson parabola, a frosted screen was used to collect the light reflected from the target, in order to estimate the laser energy absorption.

Figure 4.2 reports the main experimental results of the 2012 campaign: an enhancement of laser-target coupling for grating targets compared to simple flat targets was actually observed. Moreover, for grating targets the curve of the cut-off energy as a function of the incidence angle shows a peak at 30°, which was the expected angle of the resonance. The left panel shows data for the simple flat targets, whereas the right panel shows the results for the grating targets.

The reflected laser light (black dashed curve and hollow circles) shows a sharp dip in correspondence of the expected resonance angle for the grating targets. Indeed the integrated reflected signal plunges from $\sim$1 (arbitrary units), which is comparable with the value for simple target, down to $\sim$0.1. This suggests a strong enhancement of laser absorption.

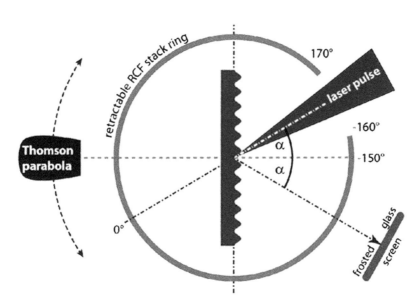

Fig. 4.1 Setup of the 2012 experimental campaign. Grating targets were irradiated at various incidence angles. A Thompson Parabola, aligned with the target normal, was used to collect the spectrum on TNSA ions. A frosted screen, imaged by a CCD, was used to acquire information on the amount of laser energy reflected by the grating. Reprinted figure with permission from Tiberio Ceccotti et al. Physical Review Letters 111, 18 [5]. Copyright 2013 by the American Physical Society

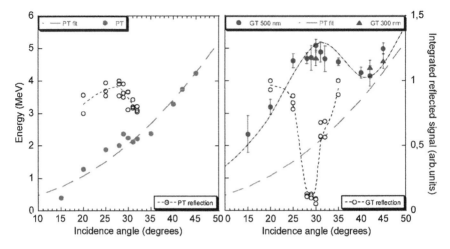

Fig. 4.2 The two graphs show the cut-off energy of accelerated protons (*filled data points*) and the reflected light signal (*empty data points*), as a function of the pulse incidence angle. Data for 20 μm thick flat targets are reported in the left frame, while the other frame show the data for 23 μm thick grating targets. Also gratings with different peak-to-valley depth were tested (*filled circles* and *triangles* correspond to 0.5 and 0.3 μm deep gratings, respectively). Reprinted figure with permission from Tiberio Ceccotti et al., Physical Review Letters 111, 18 [5]. Copyright 2013 by the American Physical Society

As far as the Thompson parabola is of concern, if compared with simple flat targets, energy spectra for ions accelerated with gratings were characterized by a significantly higher cut-off energy. In Fig. 4.2 the red points are the collected experimental cut-off energies for the flat targets (the red dashed curve is a fit), while the blue points are the analogous experimental results for grating targets (the blue dash-dotted curve is a fit). The observation of a maximum of the cut-off energies around the expected angle for resonant excitation enhancement suggests that the excitation of surface plasmons may be involved in the process. From the left frame the resonant enhancement of ion cut-off energies is evident (for pulse incidence close to 30°, cut-off energies for grating targets are approximately twice as larger as for simple flat targets. Only at large angles of incidence, gratings and flat foils behave similarly, as far as cut-off energies are of concern.

2D numerical simulations included in [5] were qualitatively in agreement with the experimental results (ALaDyn [6] PIC code was used).

As far as the study of electron emission from irradiated grating targets is of concern, so far this topic has been experimentally studied only at moderate intensities [7–9], no greater than 10^{16} W/cm^2, which is far below the intensity required to drive a relativistic surface plasmon.

4.1.2 Previous Theoretical and Numerical Investigations

Very few numerical and theoretical works deal with surface plasmons excitation at relativistic intensities.

As already mentioned in the previous subsections, some Particle-In-Cell simulations were performed to get some insights on the enhanced ion acceleration with grating targets irradiated at their resonance angle. A 2D PIC simulation of electron acceleration with SP was carried out in [10], but only at sub-relativistic intensities ($a_0 = 0.85$) and with a plane-wave laser. In a very recent work which explores electron acceleration regimes in surface plasma wave [11], possible self-injection and phase-locking of electrons at relativistic intensity in the surface plasma wave is shown with a test particle approach.

As far as the theoretical investigation of the process is of concern, in [12] a theoretical model of TNSA ion acceleration enhanced by SP excitation was proposed, with specific reference to the experiment described in Sect. 4.1.1. In [13] a tutorial treatment of surface plasmon excitation in the high field regime is provided.

4.2 Experimental Campaign at CEA-Saclay

As previously mentioned, the experimental campaign described in this chapter stemmed from the 2012 campaign on enhanced TNSA ion acceleration performed at CEA-Saclay.

In this second experimental campaign, grating targets similar to those tested in 2012 were irradiated at various incidence angles. A new experimental set-up was prepared in order to study the main properties of the electrons emitted from the target (see Sect. 4.2.1): their energy spectrum and their spatial distribution. The experimental activity took place at the Saclay Laser-matter Interaction Center (SLIC), which is part of the large research facility CEA-Saclay (Gif-sur-Yvette, France). SLIC hosts several laser systems, tailored to different scientific purposes. The experiment discussed in this thesis was performed exploiting the UHI100 laser system, which is able to deliver a 25 fs pulse with a peak power of 100 TW. UHI100, as essentially all the ultra-high intensity systems, is a Ti-Sapphire laser, which is based on the CPA scheme to amplify the pulse up to an intensity of 5×10^{19} W/cm^2. Thanks to a double plasma mirror [14, 15], the laser pulse is characterized by a very good temporal contrast (better than 10^{-10}) and adaptive optics are used to correct the pulse phase-front and to optimize the focal spot. The main properties of the laser pulse are summarized in Table 4.1. Since the targets used for the experiment were characterized by sub-wavelength structures (the peak-to-valley depth of the gratings was $\sim\lambda/4$), a very good pulse-to-prepulse contrast is critical to avoid early damages of their surface. Due to its excellent contrast, the UHI100 facility is ideal for this research activity.

Table 4.1 Main parameters of UHI100 laser system (SLIC, CEA-Saclay, Gif-sur-Yvette, France)

UHI100 laser system	
Duration	25 fs
Maximum energy	∼2.5 J
Repetition rate	10 Hz
Wavelength	∼800 nm
Maximum contrast	10^{-12}
Peak intensity on target	5×10^{19} W/cm^2
Peak power	100 TW
Technology	Ti:Sapphire—CPA system

4.2.1 Experimental Setup

Figure 4.3 shows schematically the experimental setup adopted for the whole experimental campaign. The pulse was focused on target with an off-axis f/3.75 parabola in a focal spot of $\simeq 4\,\mu$m FWHM containing ∼60 % of the total energy in the $1/e^2$ spot diameter, which lead to an average intensity of ∼5×10^{19} W/cm^2. Focal spot optimisation was performed with an adaptive optical system. P-polarization was used throughout the experiment.

A Lanex scintillating screen and an electron spectrometer were used (alternatively) to collect information on electrons emitted from the irradiated targets. A Thompson parabola aligned with the target normal and aimed at the back face of the target was used to collect ion spectra and to find the best focus condition. The diagnostics

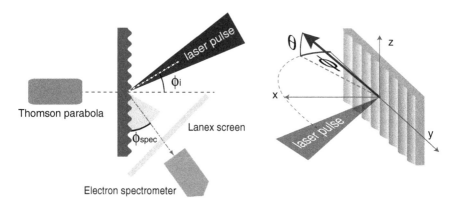

Fig. 4.3 Experimental setup of the campaign performed at CEA-Saclay facility. The left panel shows a schematic of the experiment, showing all the diagnostics used during the experiment (Thomson parabola, electron spectrometer and Lanex scintillating screen). Naming conventions for the pulse incidence angle and the angle of view of the spectrometer are shown. The panel on the *left* is a reference for the naming convention for the pulse incidence angle and the emission angles

shown in the scheme are described in detail hereunder. Some information on the grating targets and their fabrication process is provided as well.

Lanex Scintillating Screen

The Lanex scintillating screen is a commercial device (Fast screen from Carestream) which is used in the experimental setup to collect information on the spatial distribution of emitted electrons. These screens incorporate phosphors, which convert deposed energy into green light emission (through the luminescence process). An imaging system based on a CCD camera is used to record light emission from the Lanex.

Lanex screens are sensitive to x-rays and charged particles (both electrons and ions) and their response can be calibrated using conventional accelerators or known radiation sources (see for example [16]).

During the experimental activity the Lanex screen was shielded with a 3 mm Al foil, in order to filter out electrons with energy lower than $\sim$1.5 MeV ([17]).

The Lanex screen and its imaging system were calibrated with a stable known electron source (the LINAC at Laboratoire de l'Accélérateur Linéaire, Orsay, France).

Electron Spectrometer

An electron spectrometer is a device designed to collect information on the electron spectrum, aiming at its reconstruction. The spectrometer we used was specifically built for the experimental campaign on grating targets.

Essentially, the electrons enter the device through a circular aperture and are then deflected by a magnetic field onto a scintillator foil (Lanex). The scintillator is glued on a large (49, 2 × 76, 8 mm) triggered 12bit CMOS with 48 μm pixel size.

The detector window is shielded with a thin aluminium foil (a few μms), while a 2.5 mm-thick lead pin-hole is placed in front of the magnet (the hole diameter is 500 μm). The properties of the magnet and of the CMOS detector allowed a detection range from $\sim$2 up to $\sim$30 MeV.

It was possible to control remotely the position of the magnet and of the pin-hole. If the magnet and the pin-hole are removed, the device can be used in "imaging mode" to study the spatial features of the electrons coming from the target. However, with the magnet removed, the electron signal is superimposed with an intense hard x-ray background.

The electron spectrometer assembly was mounted on motorised tray which was able to change the angle ϕ_{spec} remaining aligned to the interaction center without opening the vacuum chamber.

The detector is completely shielded from the external light, being enclosed in a thick metal box. The cables (needed to power the motors and the detector itself) enter the box through a chicane. This shielding is beneficial to increase the sensitivity of the device.

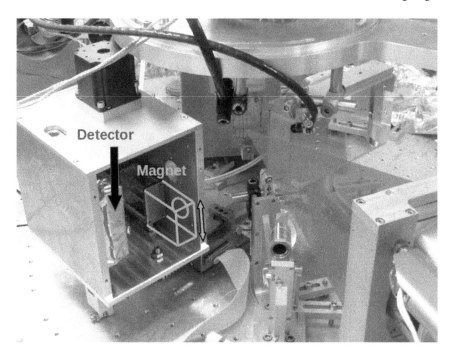

Fig. 4.4 The electron spectrometer is shown with its lateral panel removed. The magnet had been removed when the picture was taken, thus its position is highlighted in *green* (colour figure online)

Significant effort was required to allow the device to work in trigger-mode (i.e. to be switched on a few ms before the laser pulse and to be switched off a few ms later). Indeed the CMOS detector was not designed to operate in this mode by the manufacturer and custom acquisition program had to be written.

A possible issue of the spectrometer setup is that the incoming electrons produce signal passing through the lanex screen, but they also interact directly with the CMOS detector. This means that an absolute calibration of the Lanex screen is not enough to calibrate the detector. Thus a proper calibration of the whole assembly (Lanex+CMOS) is needed.

Figure 4.4 shows a side-view of the electron spectrometer, already placed in the vacuum chamber. The device is placed at $\sim$10 cm from the target assembly.

Thomson Parabola

A Thomson parabola is a diagnostic designed to collect the spectrum of accelerated ions. Its operation is based on a very simple principle: accelerated particles are deflected by known static electric and magnetic fields.

We suppose that the charged particles propagate along the $\hat{z}$ axis and that the fields are constant within a given region and lying along the $\hat{y}$ axis. Considering the deflections to be only a small correction to the trajectory, from the acceleration of charged particles in EM field $\mathbf{a} = \dfrac{q}{m}\left(\mathbf{E} + \dfrac{\mathbf{v}}{c} \times \mathbf{B}\right)$ we get the following final

4.2 Experimental Campaign at CEA-Saclay

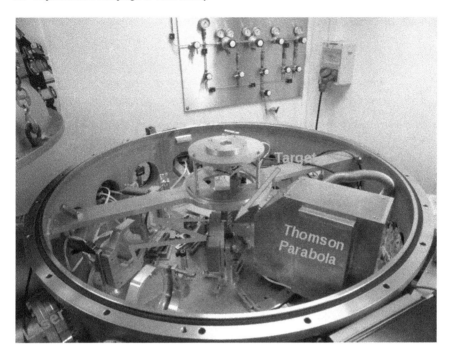

Fig. 4.5 A panoramic view of the interaction chamber. Laser path, Thomson parabola and target position are highlighted. The electron spectrometer is not visible in the picture

displacements (since in laser-plasma interaction ion velocities are usually far from being relativistic, relativistic effects are disregarded):

$$\begin{cases} \delta_x = c_1 \dfrac{q}{m} \dfrac{1}{v^2} \\ \delta_y = c_2 \dfrac{q}{m} \dfrac{1}{v} \end{cases} \tag{4.1}$$

where the constants depend on the propagation lengths and field values.

This means that $\delta_y \propto \dfrac{m}{q} \delta_x^2$. Thus, if imaged on a screen, the particles describe a parabola, whose shape depends on the charge to mass ratio. The intensity of the signal in a given point of the parabola can be used in principle to reconstruct the spectrum. If an imaging plate or a radio-chromic film is used to collect the signal the reconstruction is straightforward, since the sensitivity of these detectors is well known. A severe disadvantage of using these detectors is that they require to be processed after exposure to radiation, which could require a long time. Other detectors, such as photo-cathodes coupled with a micro-channel plate for image intensification and

Table 4.2 List of the target types used throughout the experimental campaign

Target name	Resonance angle	Groove spacing
F	–	–
G15	15°	1.35 λ
G30	30°	2 λ
G45	45°	3.41 λ

The thickness of all the targets was 10 μm, while the peak-to-valley depth of the gratings was 0.25 μm ($\approx \lambda/3$).

a CCD allow for continuous operation. However these detectors, due to inherent non-linearities, degradation over time and sensitivity to the operational conditions, cannot be calibrated straightforwardly. Deflected particles are imaged by means of a photo-cathode coupled to a micro-channel plate (MCP), which is able to intensify significantly the signal without losing its spatial structure. The electrons produced in the MCP are finally visualized on a phosphor screen (a CCD camera synchronized with the laser shot collects a the picture of the phosphor screen).

The Thomson parabola used in the setup of the experiment was characterized by a magnetic field of 0.25 T and an adjustable electric field of several 100 s kV/m. It was designed to detect the particle spectrum from a few hundreds of keV up to about 10 MeV (at higher energies the particles are only slightly deflected).

Figure 4.5 shows the Thomson Parabola (TP) assembly, already placed in the vacuum chamber. The TP was placed in order to detect protons emitted normally to the rear side of the target and it was mounted on a crescent-like rail, in order to be kept aligned with the target normal. The large tube which is visible in the picture is used to keep the CCD[2] in air, while the other parts of the device are in vacuum during normal operations.

Grating Targets

The grating targets were produced at Czech Technical University, Prague by heat embossing of Mylar™ foils using a metallic master. The target material was chosen also considering its high damage threshold for pre-pulses. Gratings with expected resonance angle at $\simeq$15°, $\simeq$30° and $\simeq$45° were produced, whose periodicities λ_g were, respectively, $\lambda_g = 1.35\lambda$, $\lambda_g = 2\lambda$ and $\lambda_g = 3.41\lambda$. The average thickness of the targets was 10 μm and the peak-to-valley depth of the grooves 0.25 μm. The peak-to-valley depth needs to be kept small with respect to the laser wavelength.

Flat foils (F) with the same average thickness were used for comparison.

The list of the targets used throughout the experiment is summarized in Table 4.2.

---

[2]CCD cameras require special cooling to operate in vacuum. The device we used was not suitable to be operated in these conditions.

4.2.2 Experimental Results

The experimental activity consisted in testing all the targets listed in Table 4.2 with various angles of incidence of the laser pulse, alternating between the two diagnostics available for electron detection. The outcome of the experimental activity is presented hereunder, starting with the results for the spatial distribution of the electron emission.

Spatial Distribution of Electron Emission

The electron emission from the front face of the target changes dramatically when gratings are used instead of flat foils. Figure 4.6 shows the spatial distribution of the electrons, as collected by the Lanex screen for two selected shots, respectively on a simple flat target and on a grating target at resonance. The emission from the flat foil is rather diffused, with a hole in correspondence of the specular reflection of the pulse. The signal reaches its maximum in a region around this hole. The ponderomotive force exerted by the reflected pulse on the electrons may be responsible for this feature. Instead, for a grating target irradiated at resonance, the emission is strongly localised at $\phi \simeq 0°$ (close to the target tangent), with an intensity ~ 10 times larger than what observed for flat foil at the same angle. The angular aperture of this collimated electron bunch is $\approx 8°$.

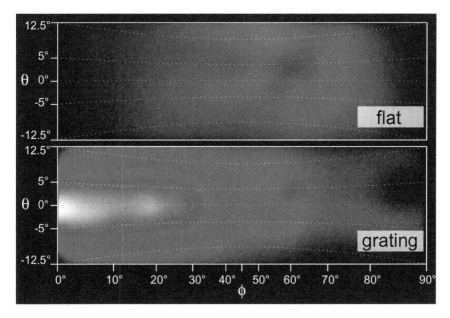

Fig. 4.6 The figure reproduces an acquisition with the Lanex screen for a simple flat target (*top*) and for grating target (*bottom*). Both targets were irradiated at $\phi_i = 30°$ pulse incidence. The Lanex screen was wrapped with an Al foil, in order to filter out electrons with energy $E \lesssim 1$ MeV. The parabolic *dashed lines* in the pictures give the local θ angle corresponding to the position on the screen

Local bending of the target or non-exact perpendicularity of the grating grooves to the plane of incidence may result in shot-to-shot fluctuations of the direction of maximum emission. Depending on the individual foil, the average angular shift in θ was in the $1°-5°$ range. An optimization of target design and alignment is foreseen to eliminate the fluctuations.

With the grating target, two minima ("holes" in the image) are observed in the directions of specular reflection and first-order diffraction of the laser pulse as if electrons were swept away by the ponderomotive force of the pulse (see Fig. 4.7, where these holes are highlighted). Since the Lanex screen and its imaging system were absolutely calibrated with a conventional acceleration, it is possible to recover an estimation of the total charge of the emitted electrons.

In the best experimental conditions (i.e. in the best series of measurements performed the same day with the grating irradiated at its resonance angle), G30 gave a total accelerated charge in the collimated bunch of 100 ± 14 pC, whereas G45 gave 130 ± 20 pC. Significantly lower charges were obtained for G15.

For comparison, with F targets irradiated at $30°$ around 60 pC were contained in the electron cloud around the specular reflection hole (typically also a rather collimated bunch was visible near the hole, containing $\sim$20 pC). When irradiated at $45°$, a charge of $\sim$250 pC was observed around the specular reflection (70 pC concentrated in the collimated bunch near the hole). It should be stressed that the angular extension of the cloud around the maximum for flat targets is much larger than the collimated bunch

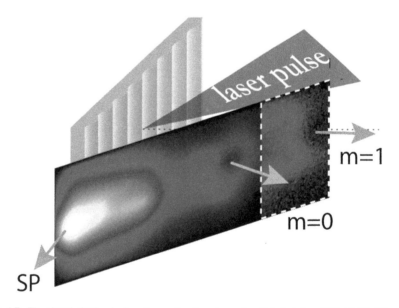

Fig. 4.7 The signal obtained when the grating target was irradiated at $\phi_i = 30°$, highlighting the position of the holes in the electron distribution. The signal of *right-hand side* of the figure (*dashed rectangle*) has been amplified 5 times to highlight the local minimum around $m = 1$

4.2 Experimental Campaign at CEA-Saclay

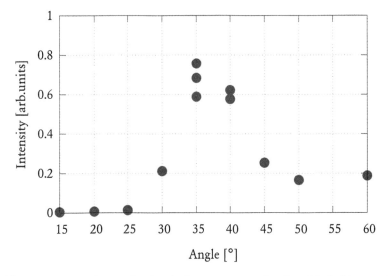

Fig. 4.8 Intensity of the electron signal on the Lanex screen as a function of the incidence angle ϕ_i. The integration of the Lanex screen signal is performed in such a way to exclude electrons emitted with $\phi > 6°$

at the target tangent for gratings. Indeed, up to 690 pC are obtained integrating the signal on the Lanex within a region of the same size for G45 irradiated at resonance.

The graph reported in Fig. 4.8 was obtained for G30 irradiated at various angles of incidence, integrating the signal on the Lanex screen corresponding to an emission close to the target tangent. A very low signal is observed for angles of incidence <30°. The signal peaks between 30° and 40° and then decreases slowly for larger angles of incidence. For most of the angles of incidence, only 1–2 points were taken, except for $\phi_i = 30°$.

Electron Spectra

Figure 4.9 shows an example of the raw results of an acquisition performed with the electron spectrometer. Knowing the properties of the device (strength of the magnet, geometry of the detector) it is possible to reconstruct the energy spectrum.

The energy spectra of the electrons emitted close to the target tangent were obtained placing the spectrometer at $\phi_{\text{spec}} = 2°$. The angle of incidence of the laser was varied from 20° to 45°. Figure 4.10 shows spectra obtained for $\phi_i \geq 30°$ as for smaller angles no clear signal above the noise level was collected. The above mentioned fluctuations of the direction of the electron beam led to a shot-to-shot variability because of the small acceptance angle of the spectrometer. Nevertheless, the most intense signals are detected only close to the resonance angle ($\sim 30°$). Moreover, spectra collected at 30° and 35° are characterized by higher maximum energies and a peculiar distribution with a dip at lower energies (3–4 MeV) and a broad peak at 5–8 MeV. On the higher energy side electrons with energy up to $\sim$20 MeV are

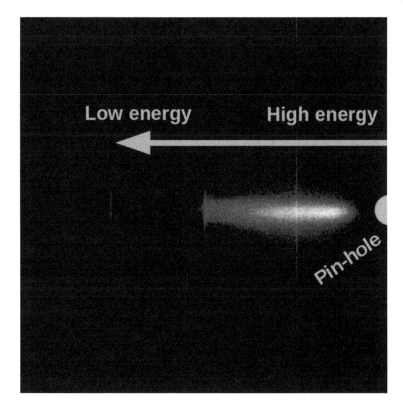

Fig. 4.9 Example of an acquired electron spectrum

detected. Whichever the pulse angle of incidence, we never observed an electron spectrum above the noise level from flat targets with $\phi_{spec} \sim 2°$. We also analysed the electron spectra obtained irradiating the target at 30° incidence and changing the position of the spectrometer in the $\phi_{spec} = 1° - 35°$ range. Despite the shot-to-shot fluctuation, the shape of the spectra remained similar in all positions of the spectrometer $\alpha_{spec} \lesssim 20°$, with a peak at 5–8 MeV. On the other hand, the intensity of the signal monotonically decreased with respect to ϕ_{spec} and the signal was visible up to $\phi_{spec} \simeq 30°$, in agreement with the integrated signal collected on the Lanex screen. We analysed the angular electron distribution for a selected case: G30 irradiated at 30° pulse incidence. Figure 4.11 shows a comparison between electron spectra obtained at different spectrometer angles. In the graphs, spectra collected with α_{spec} in the 1°–35° range are shown. Signal intensities for $\alpha_{spec} < 10°$ are, on average, significantly higher (roughly more than 2×) than for $\alpha_{spec} > 10°$. We note that, while the signal intensity shows a substantial angular dependence, the spectral shape is remarkably robust, with a peak at 4–8 MeV visible in most of the spectra, except for $\alpha_{spec} \gtrsim 20°$.

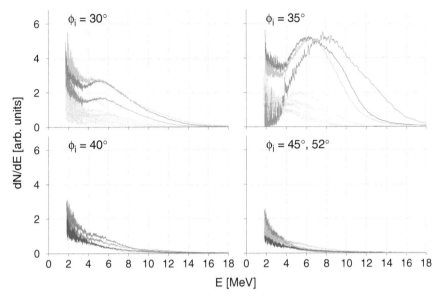

Fig. 4.10 Electron spectra collected with the detector at 2° from tangent direction, for several pulse incidence angles (from 30° to 52°). In the *upper panels*, all the collected shots are shown and a few of them are highlighted in *color*, while the others are plotted in *light grey* to show the shot to shot variations. Reprinted figure with permission from L.Fedeli, A.Sgattoni, G.Cantono, D.Garzella, F.Réau, I.Prencipe, M.Passoni, M.Raynaud, M.Kveton, J.Proska, A.Macchi, and T.Ceccotti, Physical Review Letters 116, 015001 [4]. Copyright 2016 by the American Physical Society

Holes Left in the Target

Additional signature of resonance effects in grating targets irradiated at the resonant angle is provided by the shape of the hole left in the target after the shot. These holes are significantly larger (300–500 μm) than the pulse focal spot (a few μm), their size being related to the deposed energy on target. The holes shown in Fig. 4.12 pertain, respectively, to a grating target irradiated out of resonance (e.g. G30 at 45° incidence) and a grating target irradiated at its resonant angle (e.g. G30 at 30° incidence). In the former case the holes left in the mylar foil are essentially circular, while in the latter they show a clear "tip" pointing in the direction expected for the propagation of the surface wave. This distinctive feature of the resonant condition is a clear evidence of energy propagation along the target surface for significant distances. Besides, this confirms also that the sub-μm scale structures of the grating targets survive the interaction with the pulse for enough time to produce a significant effect. The holes left in simple flat target were circular and smaller than those of the gratings, suggesting a lower efficiency in laser-plasma coupling.

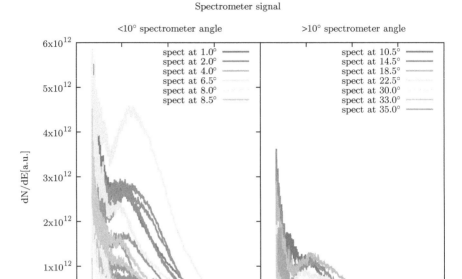

Fig. 4.11 The graph shows the electron spectra collected with the detector at several angles from tangent, for 30° pulse incidence. Only spectra collected with G30 are shown. Spectra are grouped as obtained with $\alpha_{\text{spec}} < 10°$ and $\alpha_{\text{spec}} > 10°$. The graph is reproduced from [3]

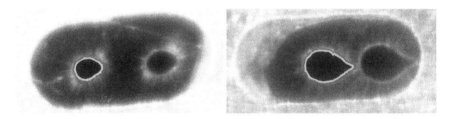

Fig. 4.12 In this picture, the holes left in the target after the shots are shown. There is a clear difference between the holes of, respectively, grating targets irradiated out of resonance and at resonance. The shape of the holes is highlighted with a *green curve*. The figure is reproduced from [3]

Thomson Parabola

Figure 4.13 is provided as an example of Thompson parabola acquisition. These data are of secondary importance for the experimental activity discussed in this chapter. Indeed, throughout the experiment, Thompson parabola was used only to find the best-focus condition. The position of the target was varied to find the highest possible cut-off energy of the accelerated protons, assuming that the best condition for ion

4.2 Experimental Campaign at CEA-Saclay 79

Fig. 4.13 Example of a Thomson parabola acquisition. *Each stripe* represents a different ion species. The *lower stripe* can be attributed to Protons, while the *other stripes* can be attributed to various ionization states of Carbon (C^{1+}, C^{2+} ...)

acceleration coincides with the best-focus (i.e. with the highest possible intensity on target). The observed ion energies were similar to those obtained in [5] (a few MeVs). The results of this diagnostic won't be discussed further.

4.3 Numerical Simulations

Numerical simulations were performed with the open-source, fully relativistic, massively parallelized Particle-In-Cell (PIC) code *piccante*. Both 3D and 2D simulations were performed. A few computationally expensive (∼100 kh) 3D simulations were performed with the aim of reproducing the experimental results, while an extensive

Table 4.3 Parameter list for 2D and 3D simulations

Simulation parameters	2D sim.	3D sim.
Simulation box	$\sim 80\lambda \times 80\lambda$	$80\lambda \times 80\lambda \times 60\lambda$
Resolution (points per λ)	102.4 (110)	[70, 51.2, 34.1]
Target density	64 n_c (120 n_c)	50 n_c
Target thickness	1 λ	1 λ
Grating depth	0.25λ	0.25λ
Grating λ_g	several [0.5λ, ∞)	2λ and ∞ (flat target)
Particles per cell (electrons)	64 (144)	50
Laser polarization	P	P
Laser a_0	5.0 (typical)	5.0
Laser waist	5 λ	5 λ
Laser duration FWHM	12 λ/c	12 λ/c
Laser incidence	several 0°−70°	30°, 35°, 40°

parametric scan was performed in 2D, in order to study the details of the physical processes at play. Table 4.3 reports the main parameters of the 2D and 3D simulations. As can be seen comparing the values quoted in Table 4.3 with the real experimental parameters, the main difference is the target thickness, which is limited to 1λ (instead of $\sim 12.5\lambda$). This thickness already requires $\sim 3 \cdot 10^{10}$ computational particles for 3D simulations, thus trying to simulate the real thickness of the targets would have required an excessive amount of computational resources. However, since we are mainly interested in physical processes taking place at the front surface of the target (SP are evanescent waves, which penetrate in the plasma only for a few skin depths), this discrepancy should be irrelevant for what happens at the front side. Another difference between the simulations and the actual experiment is the electron density of the target. A real, strongly ionized, solid target should have an electron density of 300–400 n_c, while in the simulations $n_e = 64\,n_c$ (which is still overdense, since $n_e/n_c \gg 1$). Reproducing the real density in the simulations would have required to increase the number of macro-particles per cell up to 300–400. A 2.5× increase of the spatial resolution along $\hat{x}$ would have been required to resolve the plasma skin-depth. Overall, a 40× increase of the computational cost would have been required, with presumably irrelevant improvements of the simulation.

4.3.1 2D Simulation Campaign

Almost 40 different simulations were performed to complete the 2D parametric scan. This simulation campaign was performed before the experiment at CEA-Saclay and the aim was to explore the parameter space in order to prepare the experimental activity.

4.3 Numerical Simulations

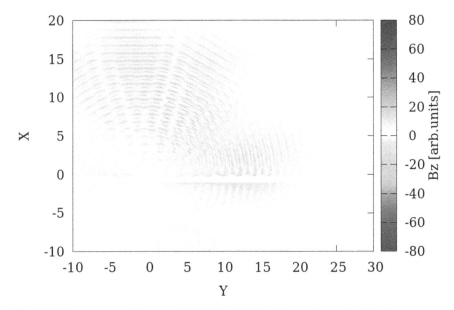

Fig. 4.14 $\hat{z}$ component of the magnetic field for a 2D simulation of an irradiated grating target

Figure 4.14 is provided to show an example of a 2D simulation of an irradiated grating target (a G30 irradiated at resonance in this case). The $\hat{z}$ component of the magnetic field is shown. The white region is the overdense grating target, in which the EM field cannot penetrate. Since the target was irradiate at its resonant angle, a surface wave propagating along the target surface can be observed.

Figure 4.15 shows a comparison of the absorption efficiency for G30, G45 and F targets, irradiated at 0°, 30°, 45°, 55° and 70°. Depending on the parameters, the absorption efficiency ranges from 5 % for flat targets irradiated at 0° up to 35–40 % for grating targets irradiated at 70° or for G30 irradiated at its resonance angle. The coupling of a laser with a corrugated target is generally enhanced with respect to the case of a simple flat target. This explains why the absorption for G30 and G45 is always greater than for F. It is important to point out that, while the absorption increases monotonically for F with the angle of incidence, G30 and G45 show an enhancement of their absorption efficiency when irradiated at their respective angles of incidence. The enhancement is significant for G30, while it is weak for G45. For large angles of incidence, both G30 and G45 behave similarly.

An enhanced absorption efficiency may reasonably lead to an enhancement of ion acceleration, since the electrons of the target are heated up to higher temperatures and are consequently able to drive a higher sheath field. As far as electron emission is of concern, 2D simulations allow to follow the acceleration process in detail. Figure 4.16 provides a sequence of snapshots from a 2D simulation of G30 irradiated at its resonance. The $\hat{z}$ component of the magnetic field is represented in grayscale, while the macro-particles with energy exceeding 5 MeV are represented as coloured

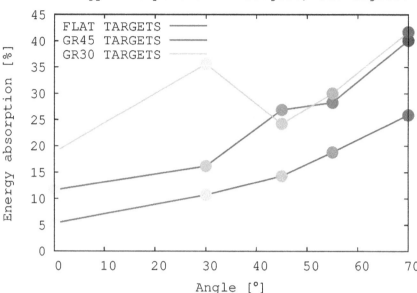

Fig. 4.15 Absorption efficiency for different targets irradiated at several angles of incidence. Data for F, G30 and G40 is reported

points. Maximum energies in the 20–30 MeV are reached and a large fraction of the high energy electrons is accelerated along the target surface, or within a small angle from the target tangent.

Looking at the trajectories of the high energy macro-particles the following acceleration mechanism can be sketched (see Fig. 4.17). An electron is extracted into the vacuum by the intense electric field component perpendicular to the surface. The extracted particle moves randomly in the EM field, until it is injected into the surface wave. The electron stays in phase with the accelerating longitudinal field for a few λ/c, reaching energies in the 20–30 MeV. Finally, at some point, the electron is expelled with a small angle from the target surface.

4.3.2 3D Simulation Campaign

Four different scenarios were tested for the 3D numerical simulation campaign: F irradiated at 30° and G30 irradiated at 30°, 35° and 40° (the parameters are reported in Table 4.3).

4.3 Numerical Simulations

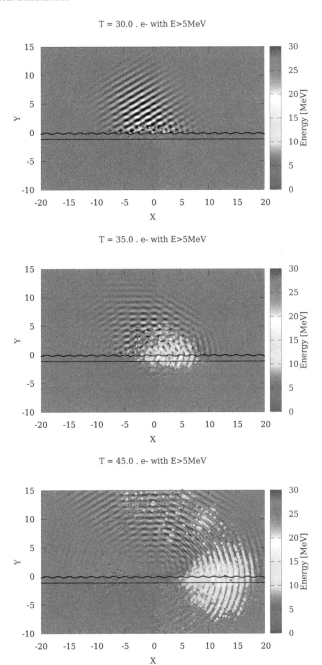

Fig. 4.16 This sequence of snapshots shows the electron acceleration process for a 2D simulation of G30 irradiated at its resonance angle. The $\hat{z}$ component of the magnetic field is represented in *grayscale*, while all the macro-particles with energy exceeding 5 MeV are represented in *color*, according to the *colorbar* provided with the graphs. The initial borders of the grating target are represented with *solid black lines*

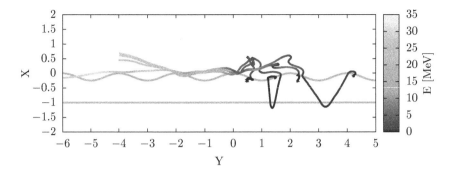

Fig. 4.17 Trajectories of a few selected high energy electrons for G30 irradiated at 30°. The grating surfaces are drawn in *gray*, while the trajectories are *coloured* according to the energy of the macro-particles

Figure 4.23 provides a snapshot of a 3D simulation of G30 irradiated at its resonance angle. The magnetic field component B_z is represented together with the isosurface corresponding to the electron density of the grating target. Since G30 was irradiated at its resonance angle, a surface wave propagating along the $-\hat{y}$ can be seen.

The primary goal of the 3D numerical simulations was to try to reproduce the experimental results, in order to shed light on the details of the acceleration process. Consequently, synthetic diagnostics (i.e. synthetic energy spectra, synthetic Lanex ...) should be produced from the 3D numerical simulation results in order to directly compare these results with the experimental ones.

Figure 4.18 shows the simulated electron energy spectra dN/dE at $\phi_{spec} = 2°$ for the flat target irradiated at $\phi_i = 30°$ and grating targets irradiated at $\phi_i = 30°, 35°, 40°$. The signal for the flat target case is very weak compared to grating spectra and it exhibits an energy cut-off which is ∼10× lower. The energy spectrum for the grating irradiated at $\phi_i = 30°$ shows the peculiar spectral shape observed for $\phi_i = 30°, 35°$ in the experimental results (see Fig. 4.10), while for higher angles of incidence the low energy dip is not observed as for $\phi_i \geq 40°$ in the experiment. Figure 4.19 compares the spectra obtained in 2D and 3D simulations. Only the 3D simulation fully reproduces details of the spectrum such as the broad peak with low energy dip.

Figure 4.20 shows the simulated angular distribution on the Lanex screen for all the 3D simulations. In order to realize these graphs, only electron with kinetic energy greater than 2.0 MeV and within the half-space $x < x_{avg}$ were considered (x_{avg} is the average x position of the grating surface[3]). Overall, the "synthetic" Lanex diagnostic reproduces fairly well the experimental data. The emission from the F target is diffused and significantly less intense than for G30 (it is enhanced by 10× in the graph). Also the hole at the specular reflection angle is evident. As far as G30 targets are of concern, the electron emission is collimated close to the target tangent

---

[3] e.g. if the grooves have their peak at $x_1 = 0$ and their valley at $x_2 = 0.25$, $x_{avg} = 0.125$.

4.3 Numerical Simulations

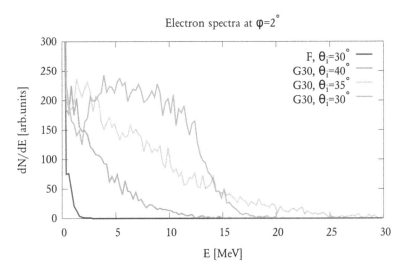

Fig. 4.18 Comparison of synthetic electron energy spectra for the 3D simulations. The spectra are calculated with $\phi_{spec} = 2°$

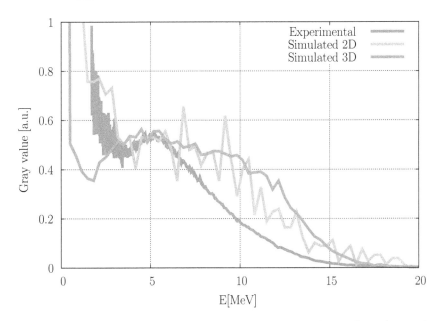

Fig. 4.19 A typical electron energy spectrum (*green*) obtained during the experiment for a grating target irradiated at $\phi_i = \phi_{res} = 30°$ (the data were collected with $\phi_{spec} = 2°$) is compared with the results of a 2D (*light blue*) and 3D (*gold*) simulation

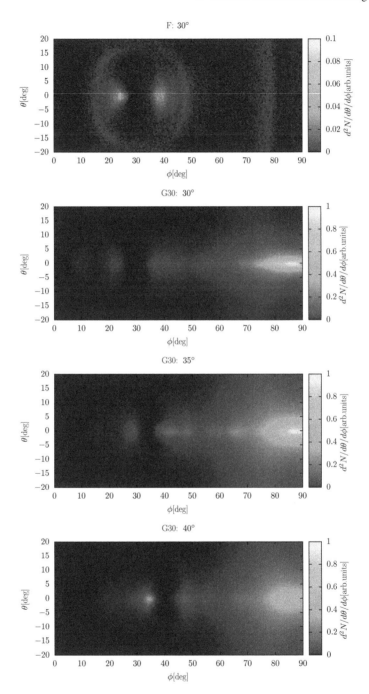

Fig. 4.20 Simulated Lanex screen (3D simulations) for F irradiated at 30° and G30 irradiated at 30°, 35°, 40°

for 30° incidence and 35° incidence, while it is rather diffused for 40° incidence. Though less evident than for F, also the angular distributions for G30 exhibit the hole at the specular reflection angle. Rather intense and collimated electron emission can be observed for G30 irradiated at 40° close to the specular reflection angle.

Figure 4.21 shows the distribution in energy and angle of the electrons $d^2N/d\phi d\phi$ for all the simulations (only electrons with $x < x_{avg}$ and $-1° < \theta < 1°$ were considered). Also in this case the results are enhanced 10× for F, due to its much weaker electron emission. It is worth to notice that for F target high energies are reached by the particles emitted close to the specular reflection (exactly at 30° the phase-space is empty, the maxima are at ∼40° and ∼20° from target tangent). However the population of particles accelerated in this way is much smaller than particles accelerated along G30 tangent. Emission for G30 irradiated at 30° is peculiar, with a significant population accelerated along target tangent and with an energy peaked around 5 MeV. The hole at lower energies is also visible. Electron emission at 35° is characterized by higher energies at the target tangent but a spectrum with a population at lower energies much higher than at higher energies. In all the cases for G30, the hole in the phase-space at the specular reflection angle is visible (Figs. 4.22 and 4.23).

The 3D simulation also shows a correlation between electron energy and emission angle. Electrons at energies lower than the peak value are emitted at some angle with respect to SP propagation direction, so that integrating over the whole range of θ the spectrum resembles the one observed in the 2D case. This is consistent with interpreting the fluctuations in the energy spectra (Fig. 4.10) as related to those in the electron beam direction.

4.4 Theory of Surface Plasmon Acceleration

Here a simple model of electron acceleration with surface plasmons is illustrated ([3, 4, 18]). We consider a dense plasma with a step-boundary, described by the following electron density function (n.b. vector **x** has components (x, y, z)).

$$n_e(\mathbf{x}) = n_{e0}\Theta(x) \tag{4.2}$$

We will write now an expression for the electromagnetic field of a Surface Wave travelling along $\hat{y}$ at the vacuum-plasma boundary. In non-relativistic theory, the dispersion relation for a surface wave in these conditions is:

$$k^2 = \frac{\omega^2}{c^2} \frac{\omega_p^2 - \omega^2}{\omega_p^2 - 2\omega^2} \tag{4.3}$$

A complete theory in the relativistic case is not available, thus we will rely on the non-relativistic theory and we will compare the theoretical model with the experimental results.

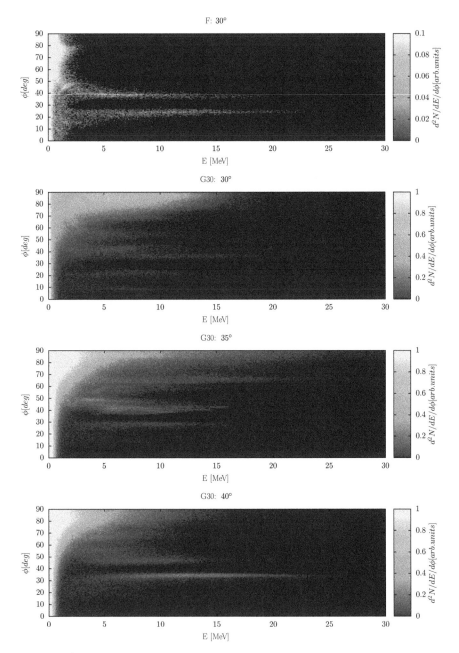

Fig. 4.21 $d^2N/dEd\phi$ from 3D simulations for F irradiated at 30° and G30 irradiated at 30°, 35°, 40°

4.4 Theory of Surface Plasmon Acceleration

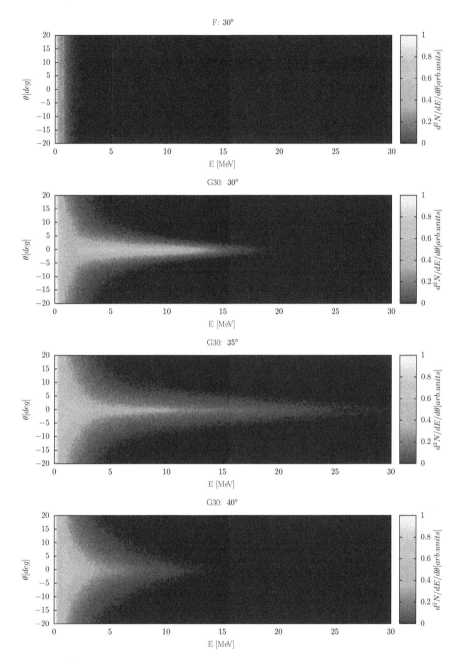

Fig. 4.22 $d^2N/dEd\theta$ from 3D simulations for F irradiated at 30° and G30 irradiated at 30°, 35°, 40°

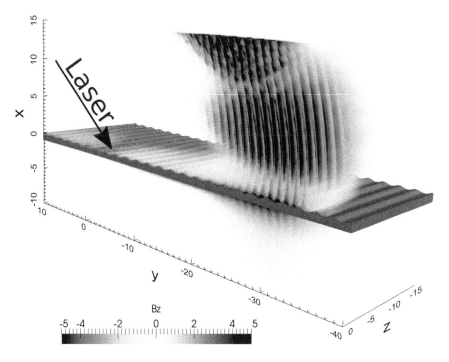

Fig. 4.23 Snapshot of a 3D simulation of G30 irradiated at resonance. The B_z component of the magnetic field and the grating surface are depicted. The simulation box was cut in half along y to ease the visualization (colour figure online)

Using Eq. 4.3, we can write the each component of the SW's electromagnetic field as a $f(\mathbf{x}) = \tilde{f}(x) \exp i(k(\omega)y - \omega t)$:

$$\begin{cases} E_x(\mathbf{x}) = -ikE_0 \left[\Theta(-x) \dfrac{e^{+q_< x}}{q_<} - \Theta(x) \dfrac{e^{-q_> x}}{q_>} \right] \exp i(k(\omega)y - \omega t) \\ E_y(\mathbf{x}) = E_0 \left[\Theta(-x) e^{+q_< x} + \Theta(x) e^{-q_> x} \right] \exp i(k(\omega)y - \omega t) \\ B_z(\mathbf{x}) = \dfrac{i\omega/c}{q_<} \left[\Theta(-x) e^{+q_< x} + \Theta(x) e^{-q_> x} \right] \exp i(k(\omega)y - \omega t) \end{cases} \quad (4.4)$$

where $\alpha = \omega_p^2/\omega^2$, $q_> = k\sqrt{\alpha - 1}$ and $q_< = k/\sqrt{\alpha - 1}$. Field components not listed in 4.4 are equal to 0.

We can now transform to a reference frame co-moving with the surface wave (i.e. a reference frame moving with a velocity $v = v_p$ with respect to the laboratory frame along $\hat{y}$, where $v_p = k/\omega$ is the phase velocity of the SW).

4.4 Theory of Surface Plasmon Acceleration

Defining β is such a way that $v_p = \beta c$ and $\gamma = 1/\sqrt{1-\beta^2}$, electromagnetic fields should be transformed as follows:

$$\begin{cases} E'_x = \gamma [E_x + \beta B_z] \\ E'_y = E_y \\ B'_z = \gamma [B_z + \beta E_x] \end{cases} \quad (4.5)$$

Obviously, transforming to a co-moving reference frame determines that primed fields do not depend on time. Indeed, the phase $\phi = ky - \omega t$ becomes $k'y'$ in the transformed reference frame, where $k' = k/\gamma$.

Performing the transformations listed in 4.5 we get:

$$\begin{cases} E'_x = -\dfrac{-i\gamma k}{q_<} E_0 \left[\Theta(-x) \dfrac{e^{+q_< x}}{q_<} \dfrac{1}{\alpha - 1} - \Theta(x) \dfrac{e^{-q_> x}}{q_>} \right] \exp k'y' \\ E'_y = E_0 \left[\Theta(-x) e^{+q_< x} + \Theta(x) e^{-q_> x} \right] \exp k'y' \\ B'_z = \dfrac{i\omega\alpha}{kc} E_0 \Theta(x) e^{-q_> x} \exp k'y' \end{cases} \quad (4.6)$$

Equations 4.6 show that the electromagnetic field in the region $x < 0$ (outside of the plasma) is purely electrostatic. It is easy to construct a potential for $\mathbf{E}'$ in this region:

$$\Phi'(\mathbf{x}') = \frac{i\gamma}{k} E_0 e^{+q_< x} \exp k'y' \quad (4.7)$$

Since the acceleration process takes place outside of the plasma, we will concentrate only on this region.

If an electron is initially at rest in the co-moving reference frame, provided that its motion remains outside of the plasma boundary, we can describe the acceleration process as a downhill motion in the potential $U = -e\Phi(\mathbf{x})$ (see Fig. 4.24). An electron almost at rest in this reference frame corresponds to an electron moving with a velocity almost equal to v_p along $\hat{y}$. In this case, the kinetic energy acquired by the electron in the reference frame co-moving with the surface wave is:

$$W' = -e\left[\Phi(\infty) - \Phi(0)\right] = \frac{eE_{sp}\gamma}{k} \quad (4.8)$$

If the electric field of the surface plasmon is strong enough, the electron becomes relativistic and the acquired kinetic energy exceeds $m_e c^2$. If this is the case, the four-momentum of the particle can be approximated as $p'_\mu \simeq (W', W'/c, 0, 0)$. Transforming back to the reference frame of the laboratory leads to $p_\mu \simeq (\gamma W', W'/c, \gamma W'/c, 0)$. The final energy and emission angle α_e are thus given by

$$\mathcal{E}_f \simeq \frac{eE_{SP}\gamma^2}{k}, \quad \tan\phi_e = \frac{p_x}{p_y} \simeq \gamma^{-1}. \quad (4.9)$$

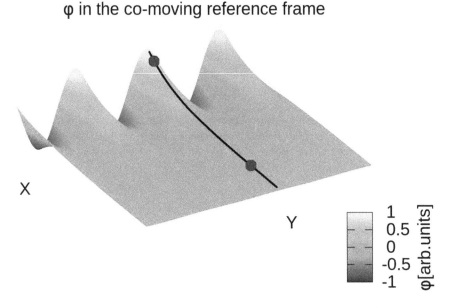

Fig. 4.24 An artistic representation of an electron accelerated by a SP. The surface shows the electrostatic potential Φ in the reference frame co-moving with the SP

Of course this calculation was done for a single particle with favourable initial conditions for acceleration up to high energies. However, this simple model already shows that strongly relativistic electrons ($\mathcal{E}_f \gg m_e c^2$) are emitted at small angles ϕ_e, i.e. close to the target surface. The acceleration length $\ell_a \equiv \mathcal{E}_f/eE_{\text{SP}} \simeq \lambda\alpha/2\pi$, showing that electrons may reach the highest energy over a few microns.

4.5 Experimental Campaign at GIST

As mentioned in the introduction, a few measurements with grating targets were realized in 2015 at a PW-class laser facility. This section is meant to give a brief overview of the preliminary results of this experimental campaign, without any in-depth analysis.

The experimental activity was carried out in 2015 at Gwangju Institute of Science and Technology (GIST, Gwangju, Rep. of Korea), where the PULSER (Petawatt Ultra-Short Laser System for Extreme science Research) laser system is available. The experiment on grating targets was performed together with the experiment of foam attached targets described in the next chapter. Only a few days were dedicated to the irradiation of the grating targets. Moreover, the characteristics of the experimental set-up at GIST allowed to test only a limited number of configurations. A first constraint is due to the large size of the vacuum chamber, which requires a few hours

4.5 Experimental Campaign at GIST

Table 4.4 Main parameters of PULSER laser system

PULSER laser system	Design parameters
Duration	30 fs
Maximum energy	44.5 J
Repetition rate	0.1 Hz
Wavelength	~800 nm
Maximum contrast	10^{-12} (@500 ps)–10^{-10} (@50 ps)
Maximum peak intensity	10^{22} W/cm^2
Maximum peak power	1.5 PW
Technology	Ti:Sapphire—CPA system

to be pumped down (compared to a few tens of minutes at CEA-Saclay). This means that manual changes of the set-up could be performed only once per day. Moreover, at GIST the detectors were not motorized, which further limited the number of testable configurations during a given experimental session.

4.5.1 Laser System

The PULSER laser system is a PetaWatt-class laser. A detailed description of the system can be found in [19, 20]. In Table 4.4 the maximum theoretical performances of the laser facility are reported. During the experimental campaign described here the laser was operated at lower pulse energies due to technical issues of the amplifier chain.

4.5.2 Experimental Set-Up

The experimental set-up is conceptually similar to that adopted for the experiment performed in France. Grating targets and flat targets were irradiated at 30° pulse incidence (their resonance incidence) and detectors to collect information of the electron spatial distribution and on the energy spectrum were used. Due to technical issues, the intensity on target was limited to a fraction of the maximum theoretical intensity: $\sim 5 \cdot 10^{20}$ W/cm^2 (essentially an order of magnitude higher than the intensity at CEA-Saclay).

The same Lanex screen used for the experiment at CEA-Saclay and the imaging system were brought to the GIST laboratory, in order to exploit their absolute calibration with the accelerator. Another (more performant) imaging device was also used to image the Lanex screen, in order to achieve a better reconstruction of the spatial distribution of the electron emission.

As far as the electron spectrometer is of concern, an Imaging Plate based detector was used.

The targets were essentially identical to those used at CEA-Saclay: simple Mylar foils and Mylar G30. In addition to these targets, a thin Al foil was also tested.

4.5.3 Preliminary Results

Two main results were obtained in the short experimental campaign. First of all the emission of collimated electron bunches at the target tangent was observed when Mylar grating targets were irradiated at their resonance angle, while no collimated bunch was recorded for simple flat Mylar target. Moreover, a peaked electron energy spectrum was collected placing the spectrometer close to the target tangent for a grating target, showing features similar to those observed in CEA-Saclay. Figure 4.25 shows the signal collected with the Lanex screen for a grating target. The collimated electron bunch at the target tangent can be observed clearly.

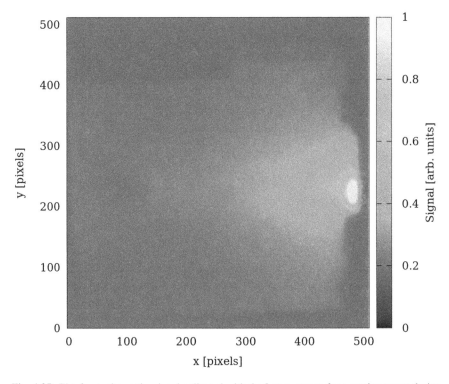

Fig. 4.25 The figure shows the signal collected with the Lanex screen for a grating target during the experimental campaign at the GIST PW-class laser facility. The collimated electron bunch close to the target tangent can be clearly seen

Irradiating metal surfaces led to an enhanced emission along the target tangent, though non collimated in θ. The exact physical process which causes this enhancement (not observed with simple Mylar targets) is still unclear. The effect might be studied further in the future.

4.6 Conclusions

The main interest of the work presented in this manuscript consists in providing compelling evidence for the excitation and propagation of surface plasmons of relativistic intensity. It is worth to stress that so far a reliable theory of relativistic SP is not available and the existence of plasmonics effects at relativistic intensities was far from being guaranteed. The results presented in this chapter might be of interest both for the plasmonics community and for the high intensity laser-plasma interaction community. Indeed, they could open the way for the extension of surface plasmon theory and possibly other plasmonic schemes into the ultra-high intensity regime. The encouraging preliminary results obtained with a PW-classe laser suggest that plasmonic effects can be observed and studied at very high laser intensities.

Besides its theoretical interest, the experimental scheme presented in this chapter might provide an attractive pulsed electron source. Intense laser interaction with grating targets provides, indeed, ultra-short, brilliant electron bunches in the few MeV energy range, up to a cut-off at $\sim$20 MeV. Electrons in this energy range are of interest for several applications, such as imaging of ultra-fast processes with electron diffraction (i.e. Ultra-fast Electron Diffraction) [21–23] or photo-neutron generation (recently laser-based photoneutron sources were proven to reach very high peak flux intensities [24, 25]). As far as the total accelerated charge is of concern, the measured value of $\sim$100 pC compares favourably to what can be obtained with LWFA in the 10s MeV energy range (see [2]). Electron bunches with good characteristics (high total charge, relatively narrow spectrum) and an energy below $\sim$20 MeV are very difficult to obtain using LWFA. Collimation of electrons accelerated with irradiated gratings is still poor with respect to "low-energy" LWFA and so is pointing stability. However LWFA is a mature acceleration technique, which has been continuously refined for decades, whereas electron acceleration mediated by relativistic SPs was observed for the first time in the present work. Thus it is well plausible that further developments of the technique might significantly improve the properties of electron emission.

References

1. K. Sakaue, Y. Koshiba, M. Mizugaki, M. Washio, T. Takatomi, J. Urakawa, R. Kuroda, Ultrashort electron bunch generation by an energy chirping cell attached rf gun. Phys. Rev. ST Accel. Beams **17**, 023401 (2014)

2. M. Mori, K. Kondo, Y. Mizuta, M. Kando, H. Kotaki, M. Nishiuchi, M. Kado, A.S. Pirozhkov, K. Ogura, H. Sugiyama, S.V. Bulanov, K.A. Tanaka, H. Nishimura, H. Daido, Generation of stable and low-divergence 10-MeV quasimonoenergetic electron bunch using argon gas jet. Phys. Rev. ST Accel. Beams **12**, 082801 (2009)
3. L. Fedeli, A. Sgattoni, G. Cantono, I. Prencipe, M. Passoni, O. Klimo, J. Proska, A. Macchi, T. Ceccotti, Enhanced electron acceleration via ultra-intense laser interaction with structured targets. Proc. SPIE **9514**, 95140R–95140R–8 (2015)
4. L. Fedeli, A. Sgattoni, G. Cantono, D. Garzella, F. Réau, I. Prencipe, M. Passoni, M. Raynaud, M. Květoň, J. Proska, A. Macchi, T. Ceccotti, Electron acceleration by relativistic surface plasmons in laser-grating interaction. Phys. Rev. Lett. **116**, 015001 (2016)
5. T. Ceccotti, V. Floquet, A. Sgattoni, A. Bigongiari, O. Klimo, M. Raynaud, C. Riconda, A. Heron, F. Baffigi, L. Labate, L.A. Gizzi, L. Vassura, J. Fuchs, M. Passoni, M. Květon, F. Novotny, M. Possolt, J. Prokůpek, J. Proška, J. Pšikal, L. Štolcová, A. Velyhan, M. Bougeard, P. D'Oliveira, O. Tcherbakoff, F. Réau, P. Martin, A. Macchi, Evidence of resonant surface-wave excitation in the relativistic regime through measurements of proton acceleration from grating targets. Phys. Rev. Lett. **111**, 185001 (2013)
6. C. Benedetti, A. Sgattoni, G. Turchetti, P. Londrillo, ALaDyn: a high-accuracy pic code for the Maxwell–Vlasov equations. IEEE Trans. Plasma Sci. **36**(4), 1790–1798 (2008)
7. G. Hu, A. Lei, W. Wang, X. Wang, L. Huang, J. Wang, Y. Xu, J. Liu, W. Yu, B. Shen, R. Li, Z. Xu, Collimated hot electron jets generated from subwavelength grating targets irradiated by intense short-pulse laser. Phys. Plasmas **17**(3), 033109 (2010)
8. G. Hu, A. Lei, J. Wang, L. Huang, W. Wang, X. Wang, Y. Xu, B. Shen, J. Liu, W. Yu, R. Li, Z. Xu, Enhanced surface acceleration of fast electrons by using subwavelength grating targets. Phys. Plasmas **17**(8), 083102 (2010)
9. S. Bagchi, P. Prem Kiran, W.-M. Wang, Z.M. Sheng, M.K. Bhuyan, M. Krishnamurthy, G. Ravindra Kumar, Surface-plasmon-enhanced MeV ions from femtosecond laser irradiated, periodically modulated surfaces. Phys. Plasmas **19**(3), 030703 (2012)
10. M. Raynaud, J. Kupersztych, C. Riconda, J.C. Adam, A. Héron, Strongly enhanced laser absorption and electron acceleration via resonant excitation of surface plasma waves. Phys. Plasmas **14**(9), 092702 (2007)
11. C. Riconda, M. Raynaud, T. Vialis, M. Grech, Simple scalings for various regimes of electron acceleration in surface plasma waves. Phys. Plasmas **22**(7), 073103 (2015)
12. C.S. Liu, V.K. Tripathi, X. Shao, T.C. Liu, Nonlinear surface plasma wave induced target normal sheath acceleration of protons. Phys. Plasmas **22**(2), 023105 (2015)
13. A. Sgattoni, L. Fedeli, G. Cantono, T. Ceccotti, A. Macchi, High field plasmonics and laser-plasma acceleration in solid targets. Plasma Phys. Control. Fusion **58**(1), 014004 (2016)
14. B. Dromey, S. Kar, M. Zepf, P. Foster, The plasma mirror: a subpicosecond optical switch for ultrahigh power lasers. Rev. Sci. Instrum. **75**(3), 645–649 (2004)
15. C. Thaury, F. Quéré, J.-P. Geindre, A. Levy, T. Ceccotti, P. Monot, M. Bougeard, F. Reau, P. d'Oliveira, P. Audebert, R. Marjoribanks, P. Martin, Plasma mirrors for ultrahigh-intensity optics. Nat. Phys. **6**, 424–429 (2008)
16. Y. Glinec, J. Faure, A. Guemnie-Tafo, V. Malka, H. Monard, J.P. Larbre, V. De Waele, J.L. Marignier, M. Mostafavi, Absolute calibration for a broad range single shot electron spectrometer. Rev. Sci. Instrum. **77**(10), 103301 (2006)
17. NIST. Stopping-power and range tables for electrons, protons, and helium ions. [Online]. Accessed 17 Sept 2015
18. A. Macchi, Personal communication
19. T.J. Yu, S.K. Lee, J.H. Sung, J.W. Yoon, T.M. Jeong, J. Lee, Generation of high-contrast, 30 fs, 1.5 PW laser pulses from chirped-pulse amplification ti:sapphire laser. Opt. Express **20**(10), 10807–10815 (2012)
20. T.M. Jeong, J. Lee, Femtosecond petawatt laser. Ann. Phys. **526**(3–4), 157–172 (2014)
21. J.B. Hastings, F.M. Rudakov, D.H. Dowell, J.F. Schmerge, J.D. Cardoza, J.M. Castro, S.M. Gierman, H.Loos, P.M. Weber, Ultrafast time-resolved electron diffraction with megavolt electron beams. Appl. Phys. Lett. **89**(18), 184109 (2006)

22. S. Tokita, S. Inoue, S. Masuno, M. Hashida, S. Sakabe, Single-shot ultrafast electron diffraction with a laser-accelerated sub-MeV electron pulse. Appl. Phys. Lett. **95**(11), 111911 (2009)
23. G. Sciaini, R.J.D. Miller, Femtosecond electron diffraction: heralding the era of atomically resolved dynamics. Rep. Prog. Phys. **74**(9), 096101 (2011)
24. I. Pomerantz, E. McCary, A.R. Meadows, A. Arefiev, A.C. Bernstein, C. Chester, J. Cortez, M.E. Donovan, G. Dyer, E.W. Gaul, D. Hamilton, D. Kuk, A.C. Lestrade, C. Wang, T. Ditmire, B.M. Hegelich, Ultrashort pulsed neutron source. Phys. Rev. Lett. **113**, 184801 (2014)
25. Y. Arikawa, M. Utsugi, A. Morace, T. Nagai, Y. Abe, S. Kojima, S. Sakata, H. Inoue, S. Fujioka, Z. Zhang, H. Chen, J. Park, J. Williams, T. Morita, Y. Sakawa, Y. Nakata, J. Kawanaka, T. Jitsuno, N. Sarukura, N. Miyanaga, H. Azechi, High-intensity neutron generation via laser-driven photonuclear reaction. Plasma Fusion Res. **10**, 2404003 (2015)

Chapter 5
Foam Targets for Enhanced Ion Acceleration

The object of the present chapter is the study (both experimental and numerical) of ion acceleration schemes with multi-layer foam attached targets.

As mentioned in the introduction and further discussed here, a significant improvement over the presently exploited schemes is required for practical applications of laser-based ion sources. In the past, schemes based on thin foils coated with a foam layer on the irradiated side gave some promising results, allowing for an enhancement of the energy of the accelerated ions. Foams consist in nano-structured assemblies of clusters, with an average density close to the critical one. The main idea is that a low density plasma should allow a better coupling with the electrons of the target, leading to a sort of enhanced TNSA scheme.

In this chapter we present the outcome of two different experimental campaigns with foam-attached targets, both carried out at the PW-class laser facility of the Center for Relativistic Lase Science (Institute for Basic Science), located at GIST (Gwangju, Republic of Korea). The experimental activity was accompanied with extensive numerical investigations performed with *piccante* particle-in-cell code.

Besides their interest as a possible route for enhanced laser-based ion acceleration, the study of laser interaction with nanostructured targets is interesting also for other reasons, ranging from achieving extremely high pressures [1] to exploring the possibility of field enhancement via the self-focusing mechanism [2].

In this chapter an initial introductory section (Sect. 5.1) states the requirements for practical applications of laser-based ion accelerators (Sect. 5.1.1) and gives an overview of the existing literature on foam attached targets (Sect. 5.1.2). The laser facility and the setup of both the experimental campaigns are presented in Sect. 5.2. Section 5.3.1 presents the results of the first campaign performed at GIST (essentially a parametric exploration of the mechanism was performed, varying the properties of both the laser pulse and the target). Section 5.3.2 describes the main result of the second experimental campaign: the effect of the laser pulse temporal length on the energy cut-off of the accelerated ions. The analysis of the results of the second experimental campaign is still ongoing, thus the results presented here should be considered preliminary. Final considerations and the outlook for the technique are discussed in Sect. 5.5.

5.1 Introduction

This section is intended to discuss the possible application of laser-driven ion sources (Sect. 5.1.1) and to frame the present work in the existing literature on foam-attached targets (Sect. 5.1.2).

5.1.1 Requirements for a Laser-Based Ion Accelerator

As anticipated in Sect. 2.2.7, the feasibility of using laser-accelerated ions for hadrontherapy is subjected to very stringent requirements. These requirements, which are far from being met even by state-of-the-art laser systems, concern: ion energies, narrowness of the energy spectrum, reproducibility of the shots, average ion current and reliability of the system. As far as the ion energies are of concern, at least $\sim$70–100 MeV are needed to treat superficial tumours (e.g. skin cancer) and much higher energies are required for tumours located deep inside the body (e.g. brain cancer), even beyond $\sim$200 MeV. The quoted thresholds are given for protons, higher energies per ion are required for heavier nuclei. Moreover, the narrowness of the energy spectrum is another critical issue. As mentioned in Sect. 2.2.7, cancer treatment with ions is interesting because the Bragg peak in the energy deposition curve allows to concentrate the tissue damage in the tumour, sparing the surrounding tissues. Energy spectra provided by laser-based accelerators are typically very broad,[1] eliminating thus the main advantage of using ions instead of electrons or photons (gamma-rays) for cancer treatment. Ions can be certainly filtered, isolating a narrow-band energy population. However, filtering reduces the total amount of accelerated charge, which is already scarce in laser based accelerators. At the moment, no more than 10^9–10^{10} ions can be accelerated with high repetition rate fs lasers, in the whole energy spectrum. However, 10^{10} particles per shot *in a small energy range* is the minimum threshold to allow medical applications [3]. At the moment, the closest attempt to a laser-based, narrow-band, proton beam source is the LIGHT project at GSI, Germany, which is able to provide bunches of $\sim$$10^9$ protons with a limited energy spread around $\sim$10 MeV (see [4]). However PHELIX, the laser system on which LIGHT is based, is a high energy Nd:YAG system, thus unsuitable for high repetition rate operations and the energy is definitely too low for any medical application.

---

[1] TNSA scheme provides very broad energy spectra, whereas RPA and shock acceleration schemes should allow for narrower spectra.

5.1 Introduction

Finally the extent of ion beam control and reproducibility achievable with traditional accelerator technology is still unmatched by laser-based systems.

In conclusion, major advancements in both laser technology and ion acceleration physics are required before laser-based hadron-therapy facilities could be conceived realistically.

A great potential exists however for several, diverse, applications in material science. Reliability requirements are generally relaxed in most of these applications, if compared to cancer treatment in human patients. Moreover, relatively low energies per nucleon and broad energy spectrum, both critical issues for laser-based hadron-therapy, are acceptable or even desirable in some cases. Some interesting applications for laser-based ion sources in material science appear to be within reach of next-generation laser facilities (the feasibility of some of these applications may be tested even with existing laser facilities).

A non-exhaustive list of possible applications of ion beams at moderate energies ($E < 100\,\text{MeV}$) could include:

- **Ion irradiation of nano-structures** Ion irradiation of nano-structures (e.g. carbon nanotubes, graphene samples) induces defects in their structures, leading to very interesting electronic properties [5, 6]. Ion bombardment is usually performed with relatively low energy ions (hundreds of KeV) and high fluencies $\sim 10^{12} - 10^{16}$ N/cm^2, using both light and heavy ion species. Irradiation with higher energies (several MeV per nucleons) and lower fluencies (down to $\sim 10^{10}$ N/cm^2) is however reported in the literature.
- **Pulsed neutron sources** Ion acceleration may be exploited for a pulsed neutron source, using a suitable converting material. Several experiments performed in recent years have reached very high neutron fluxes [7–9], as mentioned in the introduction. A possible interest is active material interrogation (useful for nuclear fuel diagnostics and detection of hidden illegal nuclear materials). Further development in laser technology may allow to reach a "supernova-like" neutron flux, possibly enabling the experiment study of r-process nucleosynthesis [10].
- **Testing of electronic components** Ion sources in the few tens of MeV per nucleon energy range are used to test electronic components for spacecraft applications. Irradiation is performed both with heavy and light ions. This technique allows to observe single event effects due to radiation damage [11].
- **Production of radioisotopes for medical applications** Very high fluxes are required for the production of radioisotopes for medical applications (reactors are presently used). This application can be foreseen only with radical innovations in laser technology (see for example [12]).

In all the aforementioned cases, the required ion energies are below 100 MeV. Current record for ion acceleration with femtosecond laser pulses is ~ 60 MeV; further development of the acceleration schemes and next-generation laser facilities should allow to reach higher energies, presumably over 100 MeV. As far as the ion energies are of concern, the requirements for several applications in material science of laser based ion accelerators are already met. Another crucial parameter is the average ion current; indeed, most of these applications would require a dose significantly higher than what could be delivered with a single shot and in some cases and an average current is needed for an extended time. Present, state-of-the-art, laser technology allows for a repetition rate of 10 Hz for 100 TW laser system and 1 Hz for PW class laser systems. Considering the repetition rate and the average charge of an accelerated ion bunch (∼1−10 nC), average ion currents up to tens of nA are already within reach of current technology. Further developments (both in laser technology and in the ion acceleration schemes) may allow for μA currents. In order to allow for high-repetition rate operations, also suitable target concepts (such as rotating targets, tape targets, gas-jets [13], liquid jets [14], droplets [15] ...) should be developed.

5.1.2 Previous Investigations with Foam Targets

Experimental investigation of multi-layer foam targets for enhanced TNSA ion acceleration was performed in the past (see [16–19]). A dedicated experimental campaign was indeed carried out at CEA-Saclay laser facility, using the UHI100 laser system (see Chap. 4 for a detailed description of this laser facility). Laser interaction with these targets was explored in the $10^{16}-10^{19}$ W/cm^2 intensity range. Targets were irradiated at 10° pulse incidence in both Low Contrast (LC, 10^8) and High Contrast (HC, 10^{12}) conditions. Foam attached targets consisted in two different types of Aluminium foils (1.5 or 10 μm thick) coupled with, respectively, 12 and 23 μm thick carbon foams with an average density of 7 ± 2 mg/cm^3 (near-critical electron density if totally ionized, sub-critical otherwise).

For intermediate intensities ($10^{16}-10^{17}$ W/cm^2) proton cut-off energies in the MeV energy range were obtained, attaining a 2−3× enhancement with respect to simple flat targets. For higher intensities the behaviour of foam attached targets approached that of simple flat targets. The enhancement at moderate energies was observed both in LC and HC configurations (with the HC configuration giving the highest cut-off energies). 2D numerical simulations performed with ALaDyn [20] code were able to reproduce qualitatively the experimental results. An in-depth numerical study of the role of the foam layer in multi-layer targets is presented in [21].

Recently very promising results were obtained at Astra-Gemini laser facility (see [22]), with targets consisting in a nanotube carbon foam attached on an ultra-thin polymer layer. The average density of the foam is close to the critical electron density. Moreover, due to the deposition technique, the scalelength of the structures of the nanotube foams is significantly smaller than the laser wavelength. This means that the foam can be considered uniform, which greatly simplifies the physics of laser-foam interaction. In [22] the main role attributed to the foam layer is the steepening and the focusing of the laser pulse, which leads to enhanced RPA when the ultra-thin foil is reached by the pulse. Both numerical and experimental evidence of this hypothesis is provided (experimental evidence is obtained collecting the temporal profile of the pulse after its passage through a free-standing foam).

Table 5.1 Main parameters of PULSER laser system

PULSER laser system	Design parameters
Duration	30 fs
Maximum energy (before compressor)	38 J (45 J in 2015)
Repetition rate	0.1 Hz
Wavelength	~800 nm
Maximum contrast	10^{-12} (@500 ps)–10^{-10} (@50 ps)
Maximum peak intensity	10^{22} W/cm^2
Maximum peak power	1.5 PW
Technology	Ti:Sapphire—CPA system

5.2 Experimental Activity

The experimental activity was carried out in two separate experimental campaigns (September 2014 and July 2015) at Gwangju Institute of Science and Technology (GIST, Gwangju, Republic of Korea), where the PULSER (Petawatt Ultra-Short Laser System for Extreme science Research) laser system is available. In Sect. 5.2 a detailed description of the experimental activity and the experimental results is provided, whereas in Sect. 5.4 the outcome of the numerical simulation campaign is presented.

5.2.1 Laser System

The PULSER laser is a PW-class laser.[2] A detailed description of the system can be found in [23, 24] and in Table 5.1 the main properties are listed. PULSER is a Ti:Sapphire CPA laser. A double plasma mirror provides high contrast pulses (3×10^{-11} contrast at 6 ps before the main pulse). Due to the geometry of the plasma mirror, after this stage the laser is S-polarized, which is rarely suitable for the experimental needs. A quarter wave-plate and a half-wave plate can be used to change the pulse polarization to C or P, leading to 13 and 18 % energy losses respectively. The maximum available energy before compressor was ~38 J in 2014 and ~45 J in 2015. However, approximately 25 % of this energy is lost in the compressor and more than 50 % is lost in the plasma mirror. The final energy on target is thus just a few Joules.

During the first experimental campaign, the shape of the focal spot was diagnosed with a high-magnification focal spot monitoring system. The focal spot diameter was less than 5 μm, with 22 % of the pulse energy concentrated within the FWHM.

---

[2]An upgrade to a 4 PW system is currently ongoing.

5.2.2 Experimental Setup

The experimental setup for both campaigns is described hereunder. Since the main focus of the experiment is the study of ion acceleration with foam-attached targets, the main diagnostics are Thomson Parabolae,[3] which are able to provide information on the energy spectra of the ions.

First Experimental Campaign
During the first experimental campaign the laser pulse was focused on target by an f/3 off-axis parabolic mirror, with an incidence angle of 30°. The final pulse energy was varied in the 1–7.4 J energy range and the intensities on-target ranged from 5×10^{19} W/cm^2 to 4.2×10^{20} W/cm^2.

As previously mentioned, the main diagnostics are Thomson Parabolae. A set of two TPs was used for this experiment. One TP was aligned with the target normal, while the other one was aligned along the laser axis. The first TP was expected to collect the most important signal, since in TNSA ions are emitted in a narrow angle (few degrees) around target tangent. TPs used in the experiment are based on micro-channel plates, which is imaged by a CCD camera. While enabling high repetition rate operations (no need to replace a single-shot imaging device), this diagnostic is very challenging to calibrate. Consequently, limited information on the total accelerated charge is provided.

Figure 5.1 shows a cartoon of the experimental setup.

In addition to the TPs, a set of three electron spectrometers was placed in the chamber. The spectrometers are based on imaging plates, which means that they can provide quantitative information on the collected energy spectra. However, only one acquisition per experimental session[4] was allowed with this diagnostic (a mechanical shutter could be opened for selected shots, allowing the electrons to enter the device). Moreover, the magnetic field of the spectrometer was probably too weak for the experimental activity. This means that the tail of the electron energy distribution could not be seen, making an estimation of the electron temperature unreliable.

Second Experimental Campaign
The experimental setup for the 2015 campaign was very similar to that of the first campaign. As far as the laser properties are of concern, 30° and 2.5° pulse incidences were tested (from now on 2.5° pulse incidence will be referred as *normal incidence*). Moreover, approximately 20 % more energy was available due to an upgrade of the laser system.

From the point of view of the diagnostics, two more TPs were added to the setup, in order to collect more information on the ion emission. However, albeit a signal was often observed on all the TPs with foam targets, the most intense signals were always collected with the TP aligned with target normal. In this manuscript only the results collected with this TP will be discussed.

---

[3] See Chap. 4 for a detailed description of these diagnostics.
[4] The vacuum chamber was opened once per day to prepare the setup.

5.2 Experimental Activity

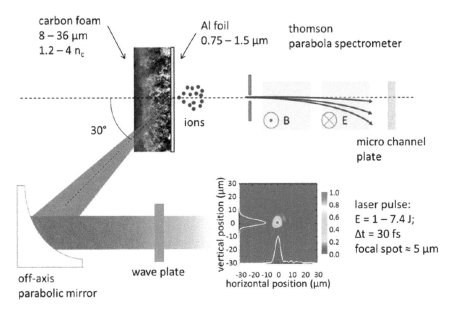

Fig. 5.1 Scheme of the experimental setup. The laser is focused on the target thanks to a parabolic mirror. The ions accelerated from the back side of the target are analyzed with a Thompson parabola. The figure is reproduced from Prencipe et al. Development of foam-based layered targets for laser-driven ion beam production, Plasma Physics and Controlled Fusion, 58 (2016) [25]. Released under a CC-BY license http://creativecommons.org/licenses/by/3.0/. Copyright IOP Publishing Ltd 2016

5.2.3 Targets

The foam attached-targets were produced at the Micro and Nanostructured Materials Laboratory at Politecnico di Milano (NanoLab), growing porous nano-structured carbon foams with Pulsed Laser Deposition [27] (PLD) technique on thin aluminium substrates, adopting a technique similar to that described in [28].

Figure 5.2 shows a typical configuration of the Pulsed Laser Deposition apparatus. A laser pulse is used to ablate a target and the ablated material is deposed on a substrate. Several parameters can be varied to control the deposition process: laser pulse properties, temperature of the substrate, pressure and type of the gas in the chamber.

For the experiment described in this chapter, a commercial Continuum Powerline II 8010 Nd:YAG laser was used to ablate a 2 in. pyrolytic graphite target. The laser is able to provide a 7 ns pulse at 10 Hz repetition rate, with a fluence of $0.8 \, J/cm^2$. Second harmonic conversion of the beam is performed, leading to a wavelength $\lambda \sim 532$ nm. The ablated species expand in the gas-filled deposition chamber and are finally collected on the substrate, forming a carbon foam layer. The gas in the chamber is Argon at 10^2 Pa, which enhances the collision frequency between the ablated targets, allowing the formation of clusters and nano-particles.

Fig. 5.2 PLD configuration (figure reproduced from [26])

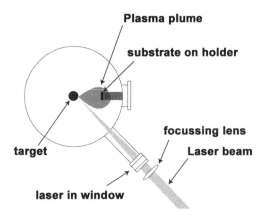

The substrate was positioned at a 45 mm distance from the target, in order to reduce inhomogeneities of the deposition process due to fluid dynamics effects. Moreover, the target was mounted on a rotating and translating stage, in order to ensure a good foam uniformity over a large area (up to a diameter of 50 mm).

The deposition parameters (e.g. laser fluence, gas type and pressure) greatly influence the physical properties of the foam (see Fig. 5.3b for an example of how

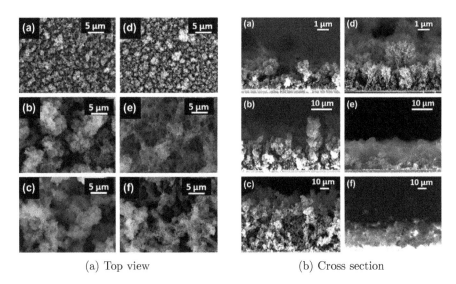

(a) Top view (b) Cross section

Fig. 5.3 These pictures are provided to show how deposition parameters affect the morphology of the foam targets. Foams shown in **a** and **d** are obtained with, respectively, Helium and Argon at 30 Pa. Foams shown in **b** and **e** are obtained with, respectively, Helium and Argon at 100 Pa. Foams shown in **c** and **f** are obtained with, respectively, Helium and Argon at 150 Pa. Reprinted from Publication Carbon, 56, A. Zani, D. Dellasega, V. Russo, M. Passoni, Ultra-low density carbon foams produced by pulsed laser deposition, 358–365 [28], Copyright 2013, with permission from Elsevier

extensively foam structures are affected by the ambient gas properties). A proper tuning of the deposition parameters allowed to reach mean mass densities as low as a few mg/cm^3 (hundreds of times less than the critical density) and thickness values in the 8–64 µm range. The foam thickness was controlled by tuning the duration of the deposition process.

Foam density was evaluated with a method based on Energy Dispersive X-ray Spectroscopy in combination with cross-section Scanning Electron Microscopy [29].

5.2.4 Experimental Plan (First Campaign)

Different target configurations were tested in the experiment: foil thickness, foam thickness and foam density were varied. Also simple aluminium targets were used for comparison. All the targets were irradiated at 30° pulse incidence. The following list summarizes all the target types used throughout the experimental activity:

- Al (0.75 µm) + foam (8.00 µm, 1.2 n_c)
- Al (0.75 µm) + foam (12.0 µm, 1.2 n_c)
- Al (0.75 µm) + foam (18.0 µm, 1.2 n_c)
- Al (0.75 µm) + foam (36.0 µm, 1.2 n_c)
- Al (0.75 µm) + foam (12.0 µm, 4.0 n_c)
- Al (1.50 µm) + foam (12.0 µm, 1.2 n_c)
- Al (1.50 µm) + foam (18.0 µm, 1.2 n_c)
- Al (0.75 µm)
- Al (1.5 µm)

Each target type was prepared in several samples. The targets were assembled with a metallic holder frame, as shown in Fig. 5.4.

5.2.5 Experimental Plan (Second Campaign)

The aim of the second experimental campaign was to further expand the parametric scan performed in the first campaign. Consequently, targets with different substrates (copper or polymer instead of aluminium) and different foam thickness were tested. The thickest substrate was a 12 µm Al foil, while the thinnest substrate was a 20 nm polymer foil. Foam layers as thick as 64 µm were tested.

Concerning the laser pulse, shots with 20 % more energy on target (at 30° pulse incidence and C-,P- polarizations) and shots at normal incidence were performed (C-polarization). In addition, the importance of pulse length was investigated. Indeed, varying the distance between the gratings in the compressor it is possible to obtain a longer and chirped laser pulse (due to non-optimal compression). Varying the distance between the gratings by hundreds of µm leads to a lengthening of the pulse up to several hundreds of femtoseconds (a proper characterization of the temporal profile of the pulse as a function of the grating spacing should still be performed).

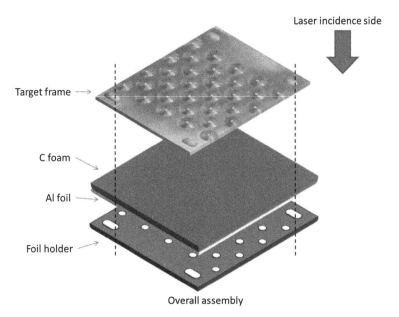

Fig. 5.4 Structure of the target assembly. The figure is reproduced from Prencipe et al. Development of foam-based layered targets for laser-driven ion beam production, Plasma Physics and Controlled Fusion, 58 (2016) [25]. Released under a CC-BY license http://creativecommons.org/licenses/by/3.0/. Copyright IOP Publishing Ltd 2016

Depending on if the gratings are moved apart or moved closer, the pulse can be positively chirped (lower frequencies first) or negatively chirped (higher frequencies first).

5.3 Experimental Results

In this section the results of both the experimental campaigns are discussed. While the results of the first campaign have been fully analysed and a detailed discussion is provided here, the results of the second campaign are still preliminary.

5.3.1 First Experimental Campaign: Enhanced Ion Acceleration

Five main results were obtained from the first experimental campaign:

- As far as ion cut-off energies are of concern, the values measured for foam-attached targets were polarization-independent (whereas a strong dependence on pulse polarization is observed for simple Al targets, as expected from the literature).

5.3 Experimental Results

- Observed cut-off energies for the smallest available foam thickness were significantly enhanced with respect to simple flat targets. The thicker available foams gave instead ion energies comparable to those obtained with simple flat targets.
- Total number of particles seems to be significantly higher for foam attached targets than for simple flat targets, at least for higher energies.
- The cut-off energies obtained with foam-attached targets were essentially independent from the substrate thickness (target thickness is instead very important for simple flat targets.
- A linear scaling of the cut-off energies was observed in the explored energy range.

Figure 5.5 shows the cut-off energies obtained for protons and C^{6+} ions for different polarizations and several foam thickness values (ranging from 0 -simple solid targets- up to 36 µm). The most important information contained in the graph is that the maximum energies are obtained for the thinnest tested foam (8 µm), while for thicker foams the maximum energies approach those obtained with the

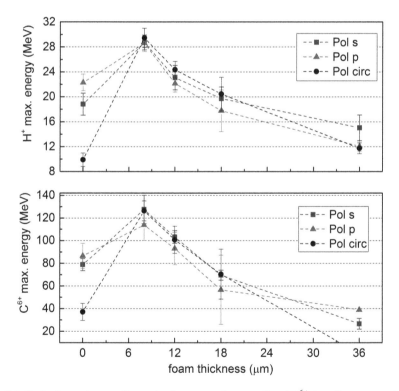

Fig. 5.5 In the figure the cut-off energies for protons (*top panel*) and C^{6+} ions (*bottom panel*) are reported for the three polarizations as a function of the foam thickness. The figure is reproduced from Prencipe et al. Development of foam-based layered targets for laser-driven ion beam production, Plasma Physics and Controlled Fusion, 58 (2016) [25]. Released under a CC-BY license http://creativecommons.org/licenses/by/3.0/. Copyright IOP Publishing Ltd 2016

flat targets. This suggests that there should be an optimal thickness d in the range $0\,\mu\text{m} < d \lesssim 8\,\mu\text{m}$. Moreover, it is evident that, while the energy cut-off is very sensitive to the laser polarization with flat targets, the experimental points for foam targets for the three polarizations are essentially superimposed. It is important to stress that the intensities on target for the three polarizations were slightly different: 4.2×10^{20} W/cm^2 for S-polarization, 3.4×10^{20} W/cm^2 for P-polarization and 3.7×10^{20} W/cm^2 for C-polarization.

Figure 5.6 shows several energy spectra collected at the maximum available laser intensity for the P-polarization. Simple flat targets and foam-attached targets are compared. Even if a proper calibration of the detector is not available, in a given energy range, the signal for foam-attached targets is significantly higher than the signal for simple flat targets.

Also the role of Al foil thickness and foam density were investigated in this experimental campaign, at least for a selected experimental condition. The right panel of Fig. 5.7 shows results obtained with S-polarized laser pulse and a foam thickness of $12\,\mu$m (if present). The graph indicates that the cut-off energy of protons is strongly dependent on target thickness for simple targets (0.75 vs 1.5 μm), while the difference between the two foam-attached targets with those substrates is negligible (well within the error bars). Moreover, increasing the foam density up to $4.3\,n_c$ did not influence the cut-off energies, since also this curve is within the error bars of the other foam attached targets.

The left panel of Fig. 5.7 shows the proton cut-off energies for various values of the foam thickness, ranging from 0 up to $36\,\mu$m as a function of the intensity on target. The acquisitions were performed with S-polarization. The data show that the cut-off energy grows linearly with the intensity. Moreover, no crossing of the curves is observed (i.e. the "optimal foam thickness" does not depend on the intensity, at least in the explored range). The figure is reproduced from [25].

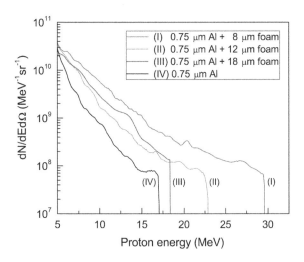

Fig. 5.6 Energy spectra of the protons for P-polarization and four different targets (simple Al foils and Al foils coupled with 8–12–18 μm foams). The figure is reproduced from Prencipe et al. Development of foam-based layered targets for laser-driven ion beam production, Plasma Physics and Controlled Fusion, 58 (2016) [25]. Released under a CC-BY license http://creativecommons.org/licenses/by/3.0/. Copyright IOP Publishing Ltd 2016

5.3 Experimental Results

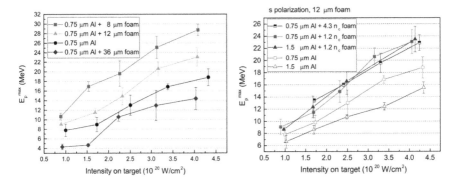

Fig. 5.7 Maximum proton energy detected as a function of the laser pulse intensity for various experimental configurations. The panel on the *left* shows experimental results for S-polarized laser pulses. *Different curves* are obtained with different values of the foam thickness. The *right panel* shows instead a comparison between different values of the substrate thickness and the foam density, showing that this parameter is unimportant for foam-attached targets. Instead, target thickness influences strongly the proton cut-off energy for simple targets. The figure is reproduced from Prencipe et al. Development of foam-based layered targets for laser-driven ion beam production, Plasma Physics and Controlled Fusion, 58 (2016) [25]. Released under a CC-BY license http://creativecommons.org/licenses/by/3.0/. Copyright IOP Publishing Ltd 2016

Despite the limits of the electron spectrometer diagnostics, the data collected with this diagnostics show that higher electron temperatures are obtained with foam-attached targets, suggesting a higher laser absorption efficiency. This supports the hypothesis that the physical process at play is essentially *enhanced TNSA*.

5.3.2 Second Experimental Campaign: Effect of Pulse Length on Ion Acceleration with Foam Targets

During the second campaign, several experimental issues with the laser system and the targets limited the number of collected data points. Moreover, diagnostics were less sensible in this experimental campaign, if a comparison is made with the set-up of the previous year, since the entrance pinholes of the Thompson Parabolae were significantly smaller. For this reason, the observed cut-off energies throughout the experimental campaign were on average lower than those observed in the 2014 campaign, even with very similar targets. In any case, a very promising effect was observed for a subset of the experimental data: varying the pulse length lead in some conditions to a significant increase (a factor of two) of the proton cut-off energy. The effect was observed in particular for a target consisting in a thin polymer substrate coated with $32\,\mu m$ neat-critical foam, irradiated at normal incidence with a C-polarized laser pulse and for another thin polymer substrate coated with $8\,\mu m$

neat-critical foam irradiated with a P-polarized pulse. However only a very limited number of experimental points was acquired, thus the evidence is non-conclusive.

Another interesting feature is the reduced influence of the substrate thickness on the cut-off energies of the ions for foam-attached targets. Indeed, among targets coated with 8 µm foam, the cut-off energy of the protons for a 12 µm thick substrate was only 30 % lower than the cut-off energy obtained with a 0.75 µm thick substrate.

More statistics is needed to confirm the observed phenomena and, eventually, to explore their features in detail. For instance, it would be interesting to understand if the enhancement of the ion energies obtained changing the position of the gratings in the compressor is related to the sign of the chirp or only to the pulse length. It would also be interesting to understand precisely the role of the target properties (substrate thickness, foam thickness …).

5.4 Numerical Simulations

Numerical simulations were performed with *piccante* to prepare the experimental activity and in order to gain some insights on the physical processes at play during intense laser-pulse interaction with foam-attached targets.

Both 2D simulations (see Sect. 5.4.1) and 3D (see Sect. 5.4.2) simulations were performed. All the simulations reported here were performed on FERMI BG/Q super-computer or on the Intel cluster Galileo, both hosted at CINECA (Italy).

5.4.1 2D Simulations

Numerical exploration of laser interaction with foam-attached targets is particularly difficult with 2D simulations in Cartesian geometry. For instance, one of the main expected features of foam targets is a relative insensitivity to pulse polarization. However, 2D simulations strongly discriminate between different polarizations (e.g. in 2D simulations P and S polarized pulses at normal incidence lead to significantly different results, despite being identical in reality). Moreover, it is important to stress that, in general, when TNSA acceleration processes are simulated in 2D, a significant un-physical enhancement of ion energies is to be expected. This is due to the fact that electrostatic phenomena in 2D involve infinite wire charges rather than point charges.

Though not reliable for quantitative estimations, 2D simulations are still useful to perform large parametric scans and to purse a qualitative understanding of the physical processes at play.

Foam Models
A first parametric exploration that can be performed concerns how the foam is simulated numerically. A foam target is a highly complex system, consisting in small

5.4 Numerical Simulations

(10s of nanometers), dense clusters aggregated in larger structures on the micrometer scale. Being near-critical on average, these foams are about 200–300 times less dense than solid carbon (e.g. graphite, diamonds ...). Since the density of the individual clusters is approximately that of solid carbon, the *filling factor*[5] of foam targets is $\sim 0.005-0.003$.

The simplest numerical approximation consists in simulating foam targets as a uniform density plasma with a steep boundary, neglecting completely their microscopic structure. This model will be referred as *uniform foam* in the following. In the literature, near critical plasmas are typically simulated with a uniform density box (like in [30]). This approach is certainly valid for CO_2 laser interaction with gas jets [31] or for solid-density targets structured scalelength smaller than the laser wavelength (e.g. nanotube array targets [22]). Uniform foam approximation seems reasonable also if we suppose that the target structures are washed out by the laser pre-pulse. However, for very high contrast lasers the aforementioned model may be questionable. Another possibility is to simulate the foam as a collection of small, dense, spheres, arranged randomly in space so that the average density is near-critical. This model will be called *random balls foam* in this chapter. The *random balls foam* has the virtue of reproducing the interaction with high density small clusters, rather than with a uniform plasma. However, larger scale structuring of the target (the formation of μm sized "trees") is completely neglected. *Diffusion Limited Aggregation (DLA) foam* model tries to reproduce the main physical features of foam targets. The foam is still simulated as a collection of dense spheres; however, these spheres are arranged in tree-like macro-structures. The position of the spheres is pre-calculated according to diffusion-limited aggregation model [32, 33] (several details, both on DLA model and its numerical implementation, are reported in Sect. 5.4.3). This method reproduces qualitatively the shape of the large scale structures of the foam (see Fig. 5.3b). One drawback of *DLA foam*, which is especially serious in 2D, is that controlling the filling factor is non-trivial. Specifically for 2D simulations, it is difficult to push the filling factor below ~ 0.05 keeping the foam shape reasonable.

The aforementioned three foam models were tested in a 2D simulation campaign, for P and C pulse polarization and for two incidence angles per model (normal incidence and 30° pulse incidence, as in the experimental activity). In Table 5.2 the parameters used for this 2D parametric exploration are listed; parameters for different foam models are listed separately.

The subsequent six pages show some results of the simulation campaign. Each page is dedicated to a single simulation case. Electron density and B_z (perpendicular to the plane) field component are shown for three different time-steps. $\lambda = 0.8\,\mu m$ is considered to express time and lengths in physical units. Figures 5.8, 5.9 and 5.10 show respectively uniform foam, random balls foam and DLA foam for 0° pulse incidence, while Figs. 5.11, 5.12 and 5.13 show the same foam-attached targets irradiated at 30°.

---

[5]The filling factor is the volume fraction actually occupied by solid-density carbon.

Table 5.2 Parameter list for 2D foam simulations

Simulation parameters	2D foam simulations
Simulation box	$125\lambda \times 100\lambda$
Resolution (points per λ)	[80, 82.24]
Al foil	
Density	$64.0\,n_c$
Thickness	$1.0\,\lambda$
Particles per cell (electrons)	64
Uniform foam	
Density	$1\,n_c$
Thickness	$1.0\,\lambda$
Particles per cell (electrons)	36
Random balls foam	
Density	$1\,n_c$ (average)
Filling factor	0.01
Thickness	$10.0\,\lambda$
Particles per cell (electrons)	576
DLA foam	
Density	$1\,n_c$ (average)
Filling factor	0.084
Thickness	$10.0\,\lambda$
Particles per cell (electrons)	64
Laser polarization	P, C, S
Laser a_0	18
Laser type	$\cos^2$ plane wave
Pulse incidence	$0°, 30°$
Laser duration FWHM	$12.3\,\lambda/c$

In all cases, the foam seems to allow a very efficient energy coupling with the laser pulse, which is able to penetrate the whole foam layer up to the dense substrate, where it is reflected back.

Table 5.3 shows the cut-off energy of protons obtained for each 2D simulation. Globally, foam-attached targets seem to allow for a better coupling with the laser pulse if compared to simple flat targets, leading to significantly higher energies. Not surprisingly, this effect is particularly evident for C-polarized laser pulses, since in this case the electron heating processes are strongly suppressed for the interaction with simple flat targets. This enhancement is coherent with the results obtained in the experimental activity. However, the two polarizations lead to very different cut-off energies, which is in substantial contradiction with the experiment (albeit not surprising, considering that the simulation is 2D).

5.4 Numerical Simulations

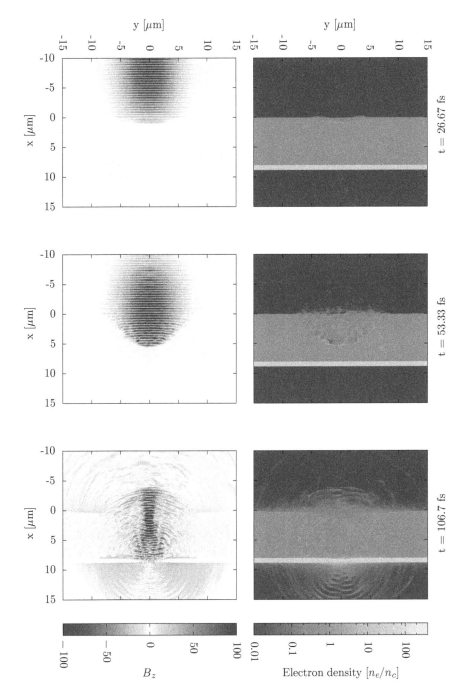

Fig. 5.8 2D simulation of laser-foam attached target interaction. Foam is simulated with a *uniform low density plasma*, pulse *incidence is* 0° (P polarization). Electron density normalized over critical density is shown in the *upper panels*. B_z field component is shown in the *lower panels*

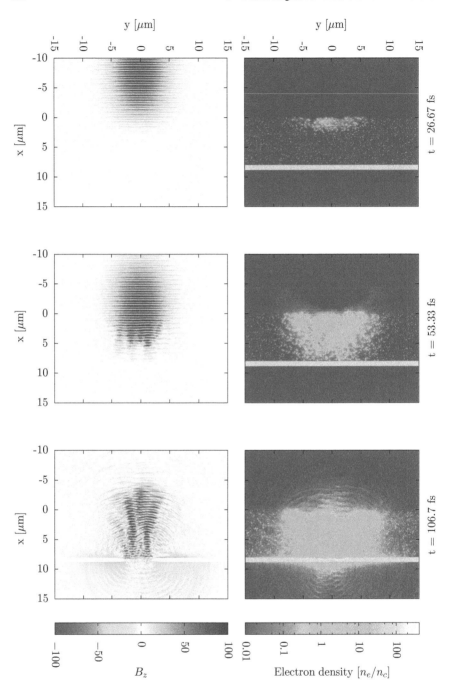

Fig. 5.9 2D simulation of laser-foam attached target interaction. Foam is simulated with a *collection of high density spheres*, pulse *incidence is* 0° (P polarization). Electron density normalized over critical density is shown in the *upper panels*. B_z field component is shown in the *lower panels*

5.4 Numerical Simulations

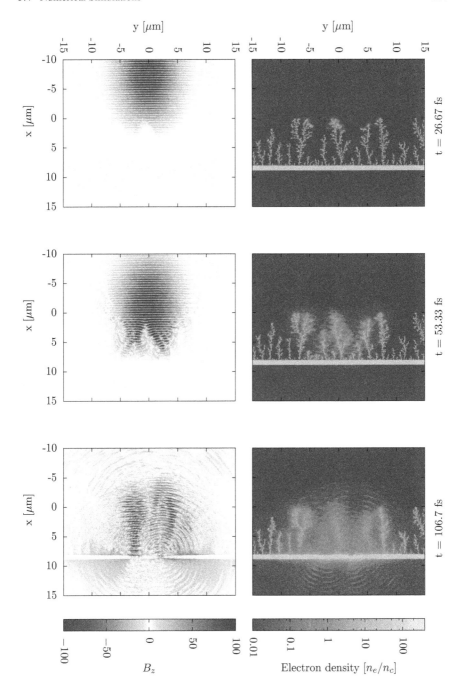

Fig. 5.10 2D simulation of laser-foam attached target interaction. Foam is simulated with a *"numerical foam"* pre-calculated with DLA, pulse *incidence is* 0° (P polarization). Electron density normalized over critical density is shown in the *upper panels*. B_z field component is shown in the *lower panels*

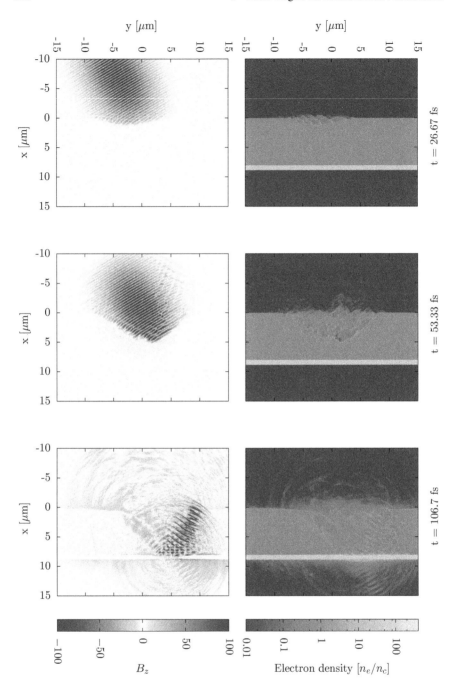

Fig. 5.11 2D simulation of laser-foam attached target interaction. Foam is simulated with a *uniform low density plasma*, pulse *incidence is* 30° (P polarization). Electron density normalized over critical density is shown in the *upper panels*. B_z field component is shown in the *lower panels*

5.4 Numerical Simulations

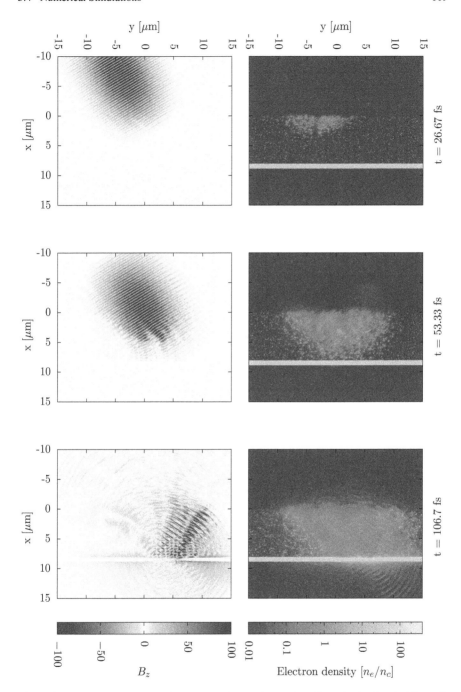

Fig. 5.12 2D simulation of laser-foam attached target interaction. Foam is simulated with a *collection of high density spheres*, pulse *incidence is* 30° (P polarization). Electron density normalized over critical density is shown in the *upper panels*. B_z field component is shown in the *lower panels*

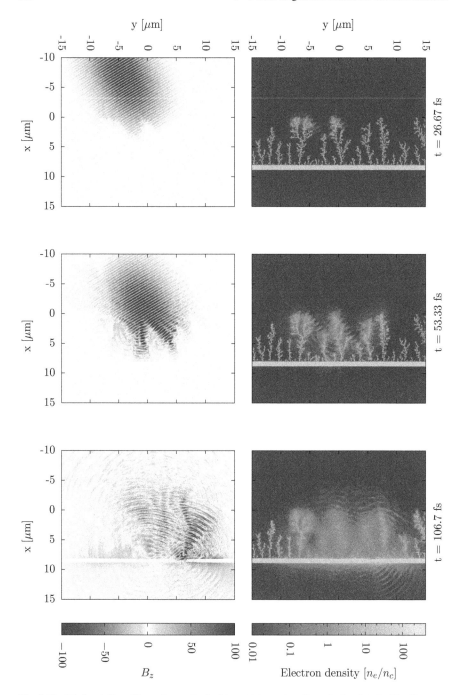

Fig. 5.13 2D simulation of laser-foam attached target interaction. Foam is simulated with a *"numerical foam" pre-calculated with DLA*, pulse *incidence is* 0° (P polarization). Electron density normalized over critical density is shown in the *upper panels*. B_z field component is shown in the *lower panels*

5.4 Numerical Simulations

Table 5.3 Synthetic results of the parametric scan performed with 2D simulations. The cut-off energy of protons is reported for each simulated configuration

#	Polarization	a_0	Incidence	Foam model	Cut-off energy (MeV)
1	P	18.0	0°	Uniform	46
2	P	18.0	0°	Random	43
3	P	18.0	0°	Cluster	47
4	P	18.0	0°	No foam	19
5	P	18.0	30°	Uniform	48
6	P	18.0	30°	Random	41
7	P	18.0	30°	Cluster	39
8	P	18.0	30°	No foam	21
9	C	12.7	0°	Uniform	25
10	C	12.7	0°	Cluster	26
11	C	12.7	0°	No foam	1.8
12	C	12.7	30°	Uniform	26
13	C	12.7	30°	Cluster	18
14	C	12.7	30°	No foam	2.1
15	P	25.5	30°	Uniform	66
16	P	25.5	30°	Cluster	61

The specific details of the foam model seem to be weakly correlated with the cut-off energy, since uniform, random and cluster foams all lead to similar energies in most cases. Also the dependence on the pulse incidence angle is rather weak, even for flat targets. This last result is in contrast with a wide body of literature, but it is important to stress that care should be taken when 2D simulations results are compared quantitatively with the experimental results, since not all physical processes are reproduced. Finally, increasing the a_0 parameter by $\sim$40 % (a doubling of the pulse intensity) resulted in a $\sim$40 % increase in the final energy, suggesting a scaling $E \propto \sqrt{I}$, typical of TNSA.

Though not being predictive, 2D simulations suggest that the enhancement of ion acceleration with foam-attached targets is due to a better coupling between the laser and the low-density foam.

5.4.2 3D Simulations

Large scale 3D–PIC simulations were performed with *piccante* code to support the interpretation of the experimental results and to investigate the effect of the foam nanostructure on the laser–target interaction process and on the acceleration performances. Only a limited number of 3D simulations was performed, due to their high computational cost (up to $\sim 1.5 \times 10^5$ CPUHours on FERMI supercomputer). Three different targets were tested in the simulations:

- a simple Al foil (referred as *ST* in the following)
- an Al foil coupled to a uniform near-critical foam
- an Al foil coupled to a "realistic" near-critical foam (referred as *DLT* in the following)

Each target was irradiated at 30° pulse incidence, with C and P polarization. Simulation parameters are summarized in Table 5.4.

The simple foil simulations and the uniform foam simulations were performed on the *Galileo* Tier-1 HPC machine, whereas the cluster simulations where performed on the *FERMI* Tier-0 HPC machine, due to larger computational requirements. Both supercomputers are housed at CINECA, Italy. Analogously to the 2D simulations, the "realistic" foam target was simulated as a collection of dense nano-spheres ($n_e/n_c \sim$ 50, 50 nm radius) arranged according to a Diffusion Limited Aggregation model. The resulting porous structure was characterized by an occupation factor of about 2 %, giving an average electron density approximately equal to n_c. The cluster foam was grown up to a maximum height of 12.5 μm and was subsequently cut at 8 μm, in order to avoid huge inhomogeneities.

Table 5.4 Parameter list for 3D simulations. *ppc* stands for *particles per cell*, *pp*λ stands for *points per* λ. The resolution is reported only for the non-stretched region of the grid

Params	Foil	Uniform	"Realistic"
Grid stretching	Yes	Yes	No
Resolution (ppλ)	[40, 20, 20]	[40, 20, 20]	[50, 20.5, 20.5]
Al n_e/n_c	40	40	40
Al thickness (μm)	0.8	0.8	0.8
Al ppc (e^-)	40	40	40
Foam n_e/n_c	–	1	1 (average)
Foam thickness (μm)	–	8	8
Foam ppc (e^-)	–	4	125
CH n_e/n_c	10	10	10
CH thickness (μm)	0.016	0.016	0.016
CH ppc (H^+)	8	8	100
Laser incidence	30°	30°	30°
Laser λ (μm)	0.8	0.8	0.8
Laser polarization	C, P	C, P	C, P
Laser a_0	18	18	18
Laser FWHM (fs)	31	31	31
Laser waist (μm)	4	4	4

5.4 Numerical Simulations

The simulations were performed on the HPC machines of CINECA, Italy: the Intel Tier-1 *Galileo* and the IBM BlueGene/Q Tier-0 *FERMI*.

The laser field amplitude had a $\cos^2$-function longitudinal profile, with full-width-half-maximum of the intensity set to 31 fs. The transverse radial profile of the laser fields are gaussian, with a waist $w_0 = 4\,\mu$m. The maximum intensity at focus is $I = 3.5 \cdot 10^{20}$ W/cm^2, corresponding to a normalized $a_0 = 18$ for linear polarization ($a_0 = (I_0 2 m_e c^3 n_c)^{1/2}$), where I_0 is the peak intensity). The laser pulse was focused at the solid foil front surface with an angle of incidence of 30°. The Al foil was represented as a uniform plasma of electrons and Al^{13+} ions with electron density $n_{Al} = 40 n_c$ and thickness $l_{Al} = 0.8\,\mu$m and it was sampled with 40 electrons per cell. Attached to the rear side of the main target a thin ($0.02\,\mu m$) low density ($n_{CH} = 10 n_c$) CH-plasma simulated the presence of a layer of contaminants on the surface of the foil. For the cases of a uniform foam, C^{6+} with density $n_{foam} = 1 n_c$ was considered and sampled 4 electron per cell. When the DLA model was considered a much higher number of electrons per cell was necessary to obtain a good sampling of the dense clusters. The clusters were spheres of radius 50 nm, the average density of the laser was $1\,n_c$ and the plasma was sampled with 125 electrons per cell.

Figure 5.14 shows the spectra of the protons of the contaminant layer emitted within 2° from the rear normal of the target, for the three simulated targets. The spectra

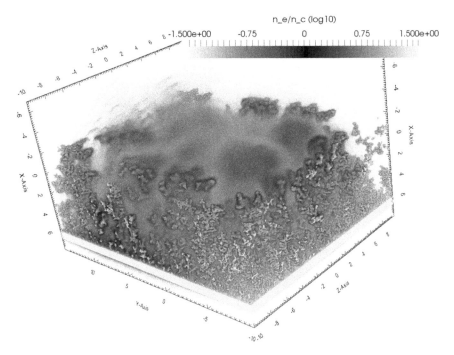

Fig. 5.14 Energy spectra from 3D simulations for protons (C- and P-polarization). An angular filter is applied, so that only particles emitted in a narrow angle around target tangent are considered

are obtained at time $t = 170$ fs after the beginning of the laser-foam interaction for both C and P polarization. Generally the cut-off energy of these spectra shows a small underestimation in comparison with the experimental data, in particular for the case of C-polarisation. The case of uniform foam irradiated with linear polarisation is an exception, since the cut-off energy is overestimated with respect to the experimental results (when compared to the case of ST, the presence of the uniform foam lead to a strong enhancement of E_p^{max}, considerably overestimating $E_p^{max-DLT}/E_p^{max-ST}$). On the other hand, when the foam layer is simulated with a DLA foam, both the difference between the two polarisations and the $E_p^{max-DLT}/E_p^{max-ST}$ are strongly reduced in very good agreement with the experimental results.

Figure 5.15 shows a sequence of snapshots of the electron density of a DLA-foam target during the interaction. Although the electron density becomes more uniform as the interaction takes place, the presence of the DLA foam strongly reduces the difference between linear and circular polarisation. Presumably, due to the coral-like structure of the DLA foam, the propagation of the pulse is strongly determined by the non uniformities of the density at the μm scalelength, rather than by the

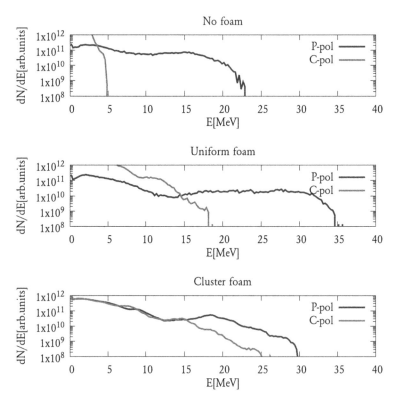

Fig. 5.15 Snapshot of a 3D simulation of laser-interaction with foam targets (the electron density of the target is shown)

5.4 Numerical Simulations

pulse-self focusing in a uniform plasma which, on the other hand, is affected by the pulse polarisation. Due to the large computational cost, the simulated DLA-foams had to be limited not to consider very small clusters below the grid mesh. The real carbon foams obtained with PLD are made of solid density clusters (100 s n_c) with a smaller radius (~10 nm) with an occupation factor about five times smaller than the simulated value of 2 %. Realistic foams with even smaller filling factors and a mesoscale structure closer to the real one might further reduce the differences between P and C polarisation in the simulations.

5.4.3 Modelling of Foam Target with Diffusion Limited Aggregation

Foam targets were modelled in *piccante* as a collection of spheres arranged in space according to the Diffusion Limited Aggregation model (see [32, 33]).

Diffusion Limited Aggregation is a very idealized model of aggregation processes. Despite its simplicity, DLA can be applied successfully to a wide variety of physical phenomena involving aggregation of small particles in larger clusters (nucleation, formation of crystals …).

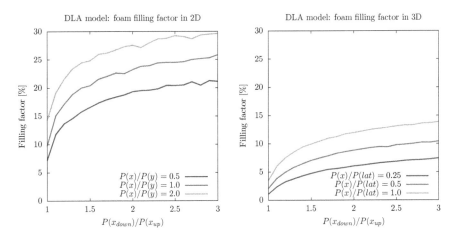

Fig. 5.16 The graphs show a parametric scan for the simulated foam targets, both in 2D and 3D. The graphs deptic the filling factors obtained varying the parameters of the "growth" process. *Three curves* are reported for three different $P(x)/P(lat)$ ratios, where $P(x)$ is the probability at each step to move up or down and $P(lat)$ is the probability to move laterally. The *curves* represent the filling factor as a function of the $P(x_{down})/P(x_{up})$ ratio, which means the ratio between the probability to move towards the target over the probability to move away from the target. If $P(x_{down})/P(x_{up}) < 1$ a single "column" is usually formed (the probability to find something to accreate on is very low, thus the highest peak is significantly favoured over the others). The graphs are obtained with several runs of the deposition algorithm

In its essence, the method is disarmingly simple. In a multidimensional lattice, individual cells can be free or occupied by one "particle". One at a time, a new particle enters the lattice and starts moving randomly (only movements to adjacent cells are allowed), until it finds a position in which at least one of the adjacent cells is occupied. Then the particle freezes in that position and a new particle enters in the lattice. The lattice can be 2D or 3D (in the literature the properties of DLA in higher dimensions have been studied, since it is a very interesting model system in statistical mechanics). Moreover, as Figs. 5.16 and 5.17 show, an interesting feature of this model is that the properties of the deposition algorithm can be tuned. Indeed, varying the probability to move in different directions (e.g. towards the target, away from the target, laterally …), very different structures with very different filling factors can be obtained. For instance, a very high probability to move towards the target leads to very "compact" foams, with high filling factors approaching the packing limit. On the contrary, a lower probability to move towards the target results in more open structures, with significantly lower filling factors. Consequently, with trials and errors, these artificial foams can be tuned in order to obtain the desired properties.

In order to avoid excessive inhomogeneities, said x_{tgt} the desired target thickness, the artificial foam was grown up to a thickness $\sim 50\%$ higher than x_{tgt} and then cut exactly at x_{tgt}.

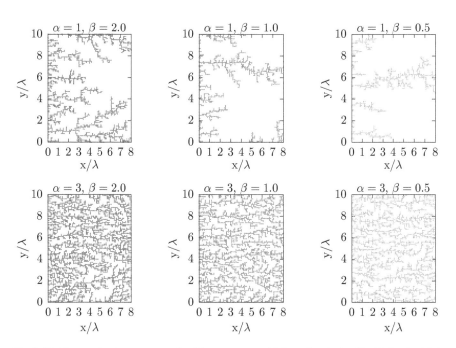

Fig. 5.17 The graphs show an example of the structures which can be obtained in 2D for the DLA deposition. α is $P(x_{down})/P(x_{up})$, while β is $P(x)/P(lat)$

5.4 Numerical Simulations

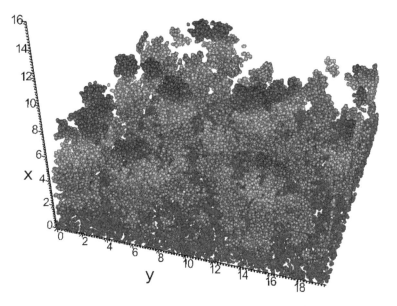

Fig. 5.18 The figure shows an example of the 3D numerical foam obtained with DLA method. *Particles' color* follows the *x* coordinates

Figure 5.18 provides a graphical representation of a 3D numerical foam target. The artificial DLA foam is "grown" with a dedicated program, specifically written for this research activity. *piccante* code can accept a file containing the coordinates of the foam particles.

5.5 Conclusions

Foam-attached targets have been studied both in experiments and numerical simulations. A significant enhancement of ion acceleration was observed in some conditions. For 30° pulse incidence, very thin near-critical foams attached to Al foil provided a strong enhancement of both the cut-off and the total number of accelerated particles,[6] if compared to simple Al targets. Moreover, irradiating at normal incidence a thick, near critical foam-attached target with a chirped (and thus longer) laser pulse led to very interesting results: a significant improvement of the cut-off energy (up to 45 MeV) was observed is some cases when the pulse length was increased up to 100–200 fs. These results suggest that foam targets are a promising scheme for ion acceleration and open the way for further experimental investigations. For instance, the very high cut-off energy obtained increasing the pulse length, even with

---

[6]Though a proper estimation of the total accelerated charge could not be performed.

a relatively modest total energy on target, is worth of further investigation, in order to unveil the exact mechanism at play.

One of the main limits of the tested foam targets was the great shot-to-shot variability, which is certainly a severe obstacle for practical applications. This variability may be attributed to the foam properties. The scalelength of the structures for a foam obtained with PLD technique is indeed at least as large as the laser wavelength. Thus the uniformity of the targets for ultra-fast laser interaction (no time for hydrodynamic expansion and "homogenization" of the target) is questionable. Another possible source of variability lies in the properties of the substrate: depending on the fabrication technique, thin aluminium foils are characterized by several holes on the surface.[7] An easier solution for this last issue is simply the adoption of more uniform substrates, such as thin copper, gold or polymer foils. Instead, a possible solution to the non-uniformity of the foam is the use of an intrinsically more uniform foam (e.g. a nano-tube based foam) as in [22]. Another possible solution is shaping the pulse in order to obtain a long pre-pulse, which should lead to a pre-expansion of the foam structures.

In conclusion, if foam attached targets will be able to provide significantly higher ion energies and/or higher particle fluxes reliably with respect to other acceleration schemes, they may enable interesting applications. Besides the possible applications in material science mentioned in the introduction of this chapter, near-critical targets can be suitable to study astrophysical scenarios (e.g. collision-less shock generation) or to reach extreme pressures [1].

References

1. M.A. Purvis, V.N. Shlyaptsev, R. Hollinger, C. Bargsten, A. Pukhov, A. Prieto, Y. Wang, B.M. Luther, L. Yin, S. Wang, J.J. Rocca, Relativistic plasma nanophotonics for ultrahigh energy density physics. Nat. Phot. **7**, 796–800 (2013)
2. C.E. Max, J. Arons, A.B. Langdon, Self-modulation and self-focusing of electromagnetic waves in plasmas. Phys. Rev. Lett. **33**, 209–212 (1974)
3. S.V. Bulanov, V.S. Khoroshkov, Feasibility of using laser ion accelerators in proton therapy. Plasma Phys. Rep. **28**(5), 453–456 (2002)
4. S. Busold, D. Schumacher, C. Brabetz, D. Jahn, F. Kroll, O. Deppert, U. Schramm, T.E. Cowan, A. Blažević, V. Bagnoud, M. Roth, Towards highest peak intensities for ultra-short MeV-range ion bunches. Sci. Rep. **5**, 12459 (2015)
5. A.V. Krasheninnikov, K. Nordlund, Ion and electron irradiation-induced effects in nanostructured materials. J. Appl. Phys. **107**(7), 071301 (2010)
6. G. Compagnini, F. Giannazzo, S. Sonde, V. Raineri, E. Rimini, Ion irradiation and defect formation in single layer graphene. Carbon **47**(14), 3201–3207 (2009)
7. I. Pomerantz, E. McCary, A.R. Meadows, A. Arefiev, A.C. Bernstein, C. Chester, J. Cortez, M.E. Donovan, G. Dyer, E.W. Gaul, D. Hamilton, D. Kuk, A.C. Lestrade, C. Wang, T. Ditmire, B.M. Hegelich, Ultrashort pulsed neutron source. Phys. Rev. Lett. **113**, 184801 (2014). Oct
8. Y. Arikawa, M. Utsugi, A. Morace, T. Nagai, Y. Abe, S. Kojima, S. Sakata, H. Inoue, S. Fujioka, Z. Zhang, H. Chen, J. Park, J. Williams, T. Morita, Y. Sakawa, Y. Nakata, J. Kawanaka, T.

---

[7]These holes are visible even with naked eye, for instance looking at a backlit target.

Jitsuno, N. Sarukura, N. Miyanaga, H. Azechi, High-intensity neutron generation via laser-driven photonuclear reaction. Plasma Fusion Res. **10**, 2404003 (2015)
9. M. Roth, D. Jung, K. Falk, N. Guler, O. Deppert, M. Devlin, A. Favalli, J. Fernandez, D. Gautier, M. Geissel, R. Haight, C.E. Hamilton, B.M. Hegelich, R.P. Johnson, F. Merrill, G. Schaumann, K. Schoenberg, M. Schollmeier, T. Shimada, T. Taddeucci, J.L. Tybo, F. Wagner, S.A. Wender, C.H. Wilde, G.A. Wurden, Bright laser-driven neutron source based on the relativistic transparency of solids. Phys. Rev. Lett. **110**, 044802 (2013). Jan
10. C. Freiburghaus, S. Rosswog, F.-K. Thielemann, r-process in neutron star mergers. Astrophys. J. Lett. **525**(2), L121 (1999)
11. J. Benitez, A. Hodgkinson, M. Johnson, T. Loew, C. Lyneis, L. Phair, Development of ion beams for space effects testing using an ECR ion source. AIP Conf. Proc. **1525**(1), 503–506 (2013)
12. M. Gerard, B. Brocklesby, T. Tajima, J. Jens Limpert, The future is fibre accelerators. Nat. Photon. **7**(4), 258–261 (2013)
13. S. Tochitsky, C. Gong, J. Pigeon, F. Fiuza, C. Joshi, He ion acceleration in near critical density plasma, in *CLEO: 2015* (Optical Society of America, 2015), p. FTh1C.4
14. J. Kim, M. Gauthier, B. Aurand, C. Curry, S. Goede, C. Goyon, J. Williams, S. Kerr, J. Ruby, A. Propp, et al., Laser-driven proton and deuteron acceleration from a pure solid-density H2/D2 cryogenic jet. Bull. Am. Phys. Soc. **60** (2015)
15. L. Di Lucchio, A.A. Andreev, P. Gibbon, Ion acceleration by intense, few-cycle laser pulses with nanodroplets. Phys. Plasmas **22**(5), 053114 (2015)
16. M. Passoni, A. Zani, A. Sgattoni, D. Dellasega, A. Macchi, I. Prencipe, V. Floquet, P. Martin, T.V. Liseykina, T. Ceccotti, Energetic ions at moderate laser intensities using foam-based multi-layered targets. Plasma Phys. Control. Fusion **56**(4), 045001 (2014)
17. Y.T. Li, Z.M. Sheng, Y.Y. Ma, Z. Jin, J. Zhang, Z.L. Chen, R. Kodama, T. Matsuoka, M. Tampo, K.A. Tanaka, T. Tsutsumi, T. Yabuuchi, K. Du, H.Q. Zhang, L. Zhang, Y.J. Tang, Demonstration of bulk acceleration of ions in ultraintense laser interactions with low-density foams. Phys. Rev. E **72**, 066404 (2005)
18. L. Willingale, S.R. Nagel, A.G.R. Thomas, C. Bellei, R.J. Clarke, A.E. Dangor, R. Heathcote, M.C. Kaluza, C. Kamperidis, S. Kneip, K. Krushelnick, N. Lopes, S.P.D. Mangles, W. Nazarov, P.M. Nilson, Z. Najmudin, Characterization of high-intensity laser propagation in the relativistic transparent regime through measurements of energetic proton beams. Phys. Rev. Lett. **102**, 125002 (2009)
19. T. Nakamura, M. Tampo, R. Kodama, S.V. Bulanov, M. Kando, Interaction of high contrast laser pulse with foam-attached target. Phys. Plasmas **17**(11), 053114 (2010)
20. C. Benedetti, A. Sgattoni, G. Turchetti, P. Londrillo, ALaDyn: a high-accuracy pic code for the Maxwell–Vlasov equations. IEEE Trans. Plasma Sci. **36**(4), 1790–1798 (2008)
21. A. Sgattoni, P. Londrillo, A. Macchi, M. Passoni, Laser ion acceleration using a solid target coupled with a low-density layer. Phys. Rev. E **85**, 036405 (2012)
22. J.H. Bin, W.J. Ma, H.Y. Wang, M.J.V. Streeter, C. Kreuzer, D. Kiefer, M. Yeung, S. Cousens, P.S. Foster, B. Dromey, X.Q. Yan, R. Ramis, J. Meyer-terVehn, M. Zepf, J. Schreiber, Ion acceleration using relativistic pulse shaping in near-critical-density plasmas. Phys. Rev. Lett. **115**, 064801 (2015)
23. T.J. Yu, S.K. Lee, J.H. Sung, J.W. Yoon, T.M. Jeong, J. Lee, Generation of high-contrast, 30 fs, 1.5 PW laser pulses from chirped-pulse amplification ti:sapphire laser. Opt. Express **20**(10), 10807–10815 (2012)
24. T.M. Jeong, J. Lee, Femtosecond petawatt laser. Ann. Phys. **526**(3–4), 157–172 (2014)
25. I. Prencipe, A. Sgattoni, D. Dellasega, L. Fedeli, L. Cialfi, I.W. Choi, I.J. Kim, K.A. Janulewicz, K.F. Kakolee, H.W. Lee, J.H. Sung, S.K. Lee, C.H. Nam, M. Passoni, Development of foam-based layered targets for laser-driven ion beam production. Plasma Phys. Control. Fusion **58**(3), 034019 (2016)
26. Wikipedia (figure released as "public domain"). Configuration PLD, 2006. [Online] Accessed 16 May 2016

27. M. Lorenz, M.S. Ramachandra Rao, 25 years of pulsed laser deposition. J. Phys. D Appl. Phys. **47**(3), 030301 (2014)
28. A. Zani, D. Dellasega, V. Russo, M. Passoni, Ultra-low density carbon foams produced by pulsed laser deposition. Carbon **56**, 358–365 (2013)
29. I. Prencipe, D. Dellasega, A. Zani, D. Rizzo, M. Passoni, Energy dispersive x-ray spectroscopy for nanostructured thin film density evaluation. Sci. Technol. Adv. Mater. **16**(2), 025007 (2015)
30. T. Nakamura, S.V. Bulanov, T.Z. Esirkepov, M. Kando, High-energy ions from near-critical density plasmas via magnetic vortex acceleration. Phys. Rev. Lett. **105**, 135002 (2010)
31. D. Haberberger, S. Tochitsky, F. Fiuza, C. Gong, R.A. Fonseca, L.O. Silva, W.B. Mori, C. Joshi, Collisionless shocks in laser-produced plasma generate monoenergetic high-energy proton beams. Nat. Phys. **8**(1), 95–99 (2012)
32. T.A. Witten, L.M. Sander, Diffusion-limited aggregation, a kinetic critical phenomenon. Phys. Rev. Lett. **47**, 1400–1403 (1981)
33. T.A. Witten, L.M. Sander, Diffusion-limited aggregation. Phys. Rev. B **27**, 5686–5697 (1983)

Chapter 6
Numerical Exploration of High Field Plasmonics in Different Scenarios

In this chapter several physical scenarios involving High Field Plasmonics effects are discussed.

In Sect. 6.1 we discuss how laser-driven Rayleigh–Taylor Instability (laser-driven RTI) [1] may arise in Radiation Pressure Acceleration (RPA) scenarios. Plasmonic field enhancement plays a central role in the development of this instability. Though essentially numerical and theoretical, this work has close ties with some early experimental observations [2]. Laser-driven RTI may play a significant role in acceleration regimes which will be attainable with next-generation laser facilities. Ion acceleration with RPA schemes is believed to become very efficient with laser intensities beyond $I > 10^{23}$ W/cm^2 [3, 4], though the development of target instabilities like laser-driven RTI should be carefully considered. Indeed, instabilities of the target may lead to its early disruption, halting the acceleration process.

In Sect. 6.2 a numerical investigation of Plasmonics effects on High Harmonic Generation (HHG) with grating targets is provided. HHG with laser-solid interaction is attractive because it could provide very intense harmonic emission [5] (in principle higher than what can be achieved with gaseous targets, provided that the laser pulse is sufficiently intense). Furthermore, using a grating as a target allows to angularly disperse the harmonics [6], which may be useful if a quasi-monochromatic source is desired. This section presents a research activity which is not yet completed, though some results are already available.

Finally, in Sect. 6.3 some speculative schemes for High Energy concentration exploiting plasmonics effects are discussed. High energy concentration is a very interesting topic in "traditional" plasmonics [7, 8]. Some possible routes to purse a similar goal in the high field regime are briefly commented. The research activity presented in this last section is very preliminary.

6.1 Rayleigh–Taylor Instability in Radiation Pressure Acceleration

The Rayleigh–Taylor instability is the classical process occurring when a heavy fluid stands over a lighter one in hydrodynamics or, equivalently, when a light fluid accelerates a heavier one [9].

In [1] we investigate numerically the role played by laser-driven RTI in Radiation Pressure Acceleration (see also [10] for a recent work on the topic). The acceleration of dense targets with the RPA scheme leads to a rippling of the interaction surface. We propose a simple model to take into account the effect of radiation pressure modulation due target rippling on the growth of the instability. We find that plasmonic enhancement of the local field when the surface rippling period is close to a laser wavelength sets the dominant scale for laser-driven RTI growth.

We provide 2D and 3D Particle-In-Cell simulation results which are in good agreement with the theoretical model. In particular, 3D simulations show the formation of stable structures with "wallpaper" symmetry, which closely resemble structures observed in early experimental investigation of RPA.

In Sect. 6.1.1 a concise introduction on Radiation Pressure Acceleration is given. In Sect. 6.1.2 our theoretical model for the calculation of laser-driven RTI growth rate is discussed in detail, while in Sect. 6.1.3 the results of the numerical simulation campaign are presented. Some conclusive remarks are finally given in Sect. 6.1.4.

6.1.1 Radiation Pressure Acceleration

Radiation Pressure Acceleration is a promising scheme for ion acceleration with laser produced plasmas. Here we follow the treatment given in [11]. More detailed information can be found in [3, 4, 12].

RPA is an ion acceleration regime which is based on radiation pressure exerted by an intense laser pulse on a target. Indeed, an electromagnetic pulse with intensity I exerts a pressure $\frac{2I}{c}$ on a perfectly reflecting mirror. For the most intense laser systems presently available, this means that the exerted pressure is 10^{16}–10^{17} Pa.[1] More in general, taking into account the reflectivity R of the target, its absorption coefficient A and its transmissivity T:

$$P = (1 + R - T)\frac{I}{c} = (2R + A)\frac{I}{c} \quad (6.1)$$

It can be shown that radiation pressure in this scenario is due to the ponderomotive force exerted on the plasma by the laser pulse.

---

[1]For comparison the pressure at the core of the Sun is $\sim 10^{17}$ Pa, according to BS05(AGS,OP) standard solar model [13].

6.1 Rayleigh–Taylor Instability in Radiation Pressure Acceleration

The ponderomotive force is exerted directly on the electrons of the target. However, their displacement generates a static electric field, which is responsible for pulling forward the ions. Thus, radiation pressure is effectively applied on the whole target.

Due to the effect of the ponderomotive force, the target surface is set in motion. If the target velocity is a non-negligible fraction of the speed of light, the radiation pressure should be calculated for a "moving mirror". Given the reflectivity $R(\omega)$ as a function of laser frequency ω, the radiation pressure exerted on the target moving with velocity $v = \beta c$ is:

$$P = P' = \frac{2I}{c} R(\omega') \frac{1-\beta}{1+\beta} \tag{6.2}$$

where primed quantities are calculated in the co-moving reference frame and the target is supposed to be non-absorbing. Radiation pressure is the same in both the laboratory frame and the co-moving frame, while $\omega' = \omega\sqrt{(1-\beta)/(1+\beta)}$ (i.e. the laser pulse is *red-shifted* in the co-moving reference frame).

In a real physical scenario, RPA is always associated with other mechanisms such as intense laser absorption and hot electron generation [14, 15] (see also Sect. 2.2.4). These phenomena can severely hinder the RPA mechanism. For instance, an intense laser pulse may break through a thin target, which leads to Coulomb Explosion [16], rather than RPA. Even if the the pulse doesn't break through the target, hot electron generation at the interaction surface may lead to TNSA, rather than RPA. Since electron heating is suppressed for a circularly polarized laser pulse irradiating a flat target at normal incidence, this is the typical configuration investigated for RPA.

Depending on the target properties, two different regimes of RPA can be individuated. If a thick target is irradiated, the laser can "dig a hole" through the plasma (Hole Boring -HB- regime), whereas a thin target can be accelerated as a whole (Light Sail -LS- regime). For the content of this chapter, only the RPA-LS regime is relevant; RPA-HB won't be discussed further (the interested reader can find more information on this topic in [3, 4]).

Light Sail Regime

For a general discussion on RPA-LS regime, the interested reader is referred to [17], here the theory will be briefly sketched.

In the RPA-LS regime, a thin, opaque foil is irradiated with a very intense laser pulse, which is responsible for accelerating the whole target. In order to model this process, we consider a rigid mirror moving along $\hat{x}$ with a velocity $\beta = \dfrac{dX}{dt}$ and a reflectivity $R(\omega)$. Said M the mass of the mirror, A its surface and $\sigma = M/A$ the areal density, we can easily write the equation of motion for the mirror under the effect of radiation pressure P:

$$\frac{d(\gamma M)}{dt} = PR(\omega\prime)A \tag{6.3}$$

replacing in the previous expression the result obtained in Eq. 6.2 we get:

$$\frac{d}{dt}(\gamma\beta) = \frac{2I(t - X/c)}{\sigma c^2} R(\omega') \frac{1 - \beta}{1 + \beta} \quad (6.4)$$

where the reflectivity $R(\omega')$ is calculated for the frequency in the reference frame co-moving with the mirror.[2] Since the mirror is moving, the intensity I is calculated at $t - X/c$, where $X(t)$ is the position of the mirror in the reference frame of the laboratory.

As noted in [17], Eq. 6.4 can be solved in a few cases, including the simple case of a perfectly reflecting mirror $R = 1$ (see also [11]). Indeed, with the substitution $w = t - X/c$ Eq. 6.4 becomes:

$$\frac{d}{dw}\left(\frac{1+\beta}{1-\beta}\right)^{1/2} = \frac{\gamma}{1-\beta}\frac{d\beta}{dw} = \frac{2I(w)}{\sigma c^2} \quad (6.5)$$

which leads to

$$\beta(w) = \frac{(1 + \mathcal{F}(w))^2 - 1}{(1 + \mathcal{F}(w))^2 + 1} \quad (6.6)$$

with $\mathcal{F}(w)$ defined as follows:

$$\mathcal{F}(w) = \frac{2\int_0^w I(w')dw'}{\sigma c^2} = \frac{2F(w)}{\sigma c^2} \quad (6.7)$$

In Eq. 6.7 $F(w)$ is the fluence of the EM wave (i.e. the total laser energy over the target surface).

Equation 6.6 allows to estimate easily the final energy per nucleon of the ions of the moving mirror, which is simply expressed as:

$$E(w) = m_p c^2 (\gamma(w) - 1) \quad (6.8)$$

evaluated for $w \to \infty$. For $w \to \infty$, $\mathcal{F}(w)$ approaches its final value $\mathcal{F}_\infty$. This leads to:

$$E_{max} = m_p c^2 \frac{\mathcal{F}_\infty^2}{2(\mathcal{F}_\infty + 1)} \quad (6.9)$$

$\mathcal{F}_\infty$ can be written as $\mathcal{F}_\infty = 2\mathcal{E}_{tot}/(Mc^2)$ ($\mathcal{E}_{tot}$ is the total laser energy, while M is the target mass). The result contained in Eq. 6.9 is particularly attractive, since it means that, for a large total fluence, the kinetic energy of the ions of the mirror scales like $\mathcal{F}_\infty$ and is thus proportional to the energy per unit surface contained in the pulse. An additional attractive feature of RPA-LS is that all the particles in the target

---

[2] This equation first appeared in [18], where spaceship propulsion with a terrestrial laser beam was envisaged as an interesting strategy for interstellar travel. In [18] the author acknowledge that an operational range of ~0.1 light year would require a coherent hard x-ray source on Earth with a surface of 1 km^2 and an x-ray mirror on the spaceship with a total surface of several km^2, making the realization of this scheme "practically impossible in the next few decades".

should be accelerated up to the same velocity. This means that, unlike TNSA, RPA-LS is expected to provide rather mono-energetic energy spectra. For a given fixed duration of the pulse the final kinetic energy is proportional to the pulse *intensity*, which implies that RPA-LS is characterized by a much more favourable scaling than TNSA. A simple estimation for present day laser systems, which are able to provide $\sim$5 J on target within $\sim$25 µm^2 gives $E_{max} \sim$100 MeV for a 20 nm thin foil of carbon ($\rho_C = 2.3$ g/cm^3). The best results obtained so far in actual experiments for ion acceleration are however significantly below the expected value from Eq. 6.9. This is essentially due to the fact that with modern day laser systems it is extremely difficult to achieve a regime of "pure" RPA-LS. Indeed, focused intensities of $\gtrsim 10^{21}$ W/cm^2 with an extremely good temporal contrast are required.

Despite being able to capture the essential features of the phenomenon, the treatment which was given for RPA-LS is oversimplified. For instance, the assumption of a perfectly rigid and perfectly reflective mirror is questionable, since the target may loose mass and become transparent, stopping the RPA-LS process. Moreover, at the aforementioned laser intensities, radiation reaction effects may become relevant. Numerical simulations are a very important tool to investigate RPA-LS regime. A numerical investigation of RPA-LS regime at laser intensities foreseen in the future can be found in [19]. Here, 3D simulations predict energies in GeV range for focused intensities of the order of 10^{23} W/cm^2. In this regime, RPA is particularly efficient because ions are promptly accelerated at relativistic velocities, which improves the energy transfer efficiency between the laser pulse and the target (when a photon is reflected by a mirror moving at relativistic velocities, it is significantly red-shifted). The role played by RR in RPA-LS is studied in detail in [20, 21].

Numerical simulations generally indicate that forthcoming 10PW laser facilities should be able to accelerate protons at least up to several hundreds of MeV. Moreover, RPA-LS should be able to provide accelerated ions with an energy spread narrower than that of TNSA.

Experimental Realization and Observation of Instabilities

Although the laser intensities needed for "pure" RPA-LS regime are not yet accessible, some experimental evidence of this physical process is already available [2, 22, 23] in which spectral features compatible with RPA-LS are observed with C-polarized laser pulses irradiating thin targets at normal incidence. An interesting outcome of these first experimental observations of RPA is the development of net-like structures in the spatial distribution of the accelerated ions. This suggests the development of an instability. Instabilities were observed also in the first numerical investigations of the phenomenon (see for instance [24]). Since an instability of the target may lead to an early onset of transparency, the process should be studied carefully.

6.1.2 Theoretical Model of Laser-Driven Rayleigh–Taylor Instability

In order to develop a theoretical model of laser-driven RTI, we consider a perfectly reflecting mirror with shallow grooves on its surface, irradiated with a monochromatic plane wave. Our aim is to write an expression for the radiation pressure on the rippled surface, in order to be able to write an equation of motion for a thin rippled target.

At first we consider a mirror which completely fills the $x > x_m(y)$ region, $x_m(y)$ setting the boundary with the vacuum. $x_m(y)$ describes the sinusoidal rippling of the mirror surface:

$$x_m = \frac{\delta}{2} \cos qy \tag{6.10}$$

where $q = \frac{2\pi}{a}$ (a is the grooves spacing) and δ is the peak-to-valley depth. We intend to pursue a perturbative approach in calculating the EM field at the mirror surface, thus we need to impose that the grooves are shallow with respect to the impinging plane wave. If ω_0 is the wave frequency and $k = \frac{\omega_0}{c}$ is the wave vector, this means:

$$k\delta \ll 1 \tag{6.11}$$

Considering deeper grooves would hinder the adoption of a perturbative approach. Moreover, we are interested in the development of the instability, starting from a perfectly flat surface. Thus the assumption of a shallow rippling is appropriate.

We then consider an impinging plane wave, with the electric field given by;

$$\mathbf{E}_i = (E_{ip}\hat{\mathbf{y}} + E_{is}\hat{\mathbf{z}}) \exp i(kx - \omega t) \tag{6.12}$$

selecting the appropriate values for E_{ip}, E_{is} allows to choose the pulse polarization: $E_{ip} = E_0$ and $E_{is} = 0$ for P polarization, $E_{ip} = 0$ and $E_{is} = E_0$ for S polarization, $E_{ip} = E_0/\sqrt{2}$ and $E_{is} = E_0/\sqrt{2}$ for C polarization. We should now solve Helmoltz equation $\nabla^2 f + k^2 f = 0$ for the electromagnetic field. The general solution can be written as:

$$\begin{cases} E_z = E_{is}e^{ikx} - E_{rs}e^{-ikx} + \sum_{l=1}^{+\infty} E_l e^{k_l x} \cos lqy \\ B_z = E_{ip}e^{ikx} + E_{rp}e^{-ikx} + \sum_{l=1}^{+\infty} B_l e^{k_l x} \cos lqy \end{cases} \tag{6.13}$$

where $k_l^2 = (l^2 q^2 - k^2)$. In writing equation system 6.13 symmetry and periodicity of the system was taken into account and the temporal dependence $e^{-i\omega_0 t}$ was omitted. Other field components can be obtained straightforwardly, using $\mathbf{E} = i\nabla \times \mathbf{B}/k$ and $\mathbf{B} = -i\nabla \times \mathbf{E}/k$.

In equation system 6.13, the first two terms represent, respectively, the incident wave and the specular reflection. The summation term would be zero for a perfectly

6.1 Rayleigh–Taylor Instability in Radiation Pressure Acceleration

flat mirror. Modes with k_l real ($ql > k$) are evanescent, thus dying off exponentially. Instead, modes with k_l imaginary ($ql < k$) are propagating modes scattered at an angle α ($\tan\alpha = lq/|k_l|$).

Assuming $E_l \sim O(\delta^l k^l)$ and using Eq. 6.11 we can truncate the summation at $l = 1$ (first-order perturbation). Thus we obtain:

$$\begin{cases} E_z = E_{is}e^{ikx} - E_{rs}e^{-ikx} + E_1 e^{k_1 x}\cos qy \\ B_z = E_{ip}e^{ikx} + E_{rp}e^{-ikx} + B_1 e^{k_1 x}\cos qy \end{cases} \quad (6.14)$$

Having chosen $l = 1$, the modes with $q > k$ are evanescent (wavelength larger than the grooves periodicity), while the modes with $q < k$ (wavelength smaller than the grooves periodicity) are propagating modes.[3]

Boundary conditions should then be imposed. The tangential component of the electric field at the target surface should be zero, since the mirror is perfect. For the same reason, the magnetic field component perpendicular to the target surface should be zero.[4] These requirements lead to:

$$\begin{cases} E_z(x = x_m(y), y) = 0 \\ \mathbf{B}\cdot\hat{\mathbf{n}}(x = x_m(y), y) = 0 \\ \mathbf{E}\times\hat{\mathbf{n}}(x = x_m(y), y) = 0 \end{cases} \quad (6.15)$$

where $\hat{\mathbf{n}}$ is the normal to the grating surface, which can be written as:

$$\hat{\mathbf{n}} = -\hat{\mathbf{x}} + x_m{'}(y)\hat{\mathbf{y}}/\sqrt{1 + x_m{'}^2} \quad (6.16)$$

For $\delta^l k^l \ll 1$, Eq. 6.16 can be expanded in a Taylor series and approximated as:

$$\hat{\mathbf{n}} \approx -\left[\hat{\mathbf{x}} + \hat{\mathbf{y}}\frac{q\delta}{2}\sin qy\right] \quad (6.17)$$

Combining Eq. 6.17 and boundary conditions in 6.15, we finally get:

---

[3] We expect a sub-wavelength grating to behave like a perfect mirror at large distances, with just some near-field perturbations (the evanescent modes).

[4] These are the standard boundary conditions at an interface between a perfect metal and vacuum. In general, the boundary conditions for EM field at an interface prescribe that the electric field should have a continuous tangential component, while, for the magnetic field, the continuous component should be the perpendicular one. However, in a perfect conductor EM field is zero, thus these components should be zero at the interface.

$$E_{rs} = E_{is} \quad E_{rp} = E_{ip}$$
$$E_{1x} = \frac{-iqk\delta}{k_1} E_{ip} \quad E_{1y} = -1k\delta E_{ip}$$
$$E_{1z} = -ik\delta E_{is} \tag{6.18}$$
$$B_{1x} = q\delta E_{is} \quad B_{1y} = k_1 \delta E_{is}$$
$$B_{1z} = \frac{k^2 \delta}{k_1} E_{ip}$$

These results imply that for P-polarized (electric field aligned with the grooves) incident waves, the rippling of the surface results in a local field enhancement in the valleys of the grooves, while for S-polarized (electric field perpendicular to the plane of the grooves) incident waves, the EM field is locally enhanced at the peaks of the grooves. It is worth to remark that for P-polarization, when $k = q$ (grooves spacing equal to the wavelength of the incident radiation), the field components E_{1y} and B_{1z} diverge. This is due to the resonant excitation of a surface wave. Although this calculation is performed for a perfect mirror, we expect that for a dense plasma a surface wave would be excited almost at the same condition.[5]

Figure 6.1 shows simulations of the reflection of an S-polarized and a P-polarized plane wave from a grating target. The averaged transverse magnetic field component is shown, confirming the theoretical prediction.

With the EM field components we can calculate the EM momentum flow via Maxwell stress tensor **T** as:
$$\mathbf{P} = \mathbf{T} \cdot \hat{\mathbf{n}} \tag{6.19}$$

Maxwell stress tensor averaged over a field oscillation period is given by:
$$T_{\alpha\beta} = \frac{1}{8\pi} \left[\Re \left(E_\alpha E_\beta^* + B_\alpha B_\beta^* \right) - \frac{1}{2} \delta_{\alpha\beta} \left(|\mathbf{E}|^2 + |\mathbf{B}|^2 \right) \right] \tag{6.20}$$

We need to evaluate this expression at the mirror surface ($x = x_m(y)$). Using the results of previous calculations we get:

$$P_x \approx \begin{cases} \dfrac{E_0^2}{4\pi} \Re[1 - k_1 \delta \cos qy] & \text{(S)} \\ \dfrac{E_0^2}{4\pi} \Re\left[1 + \dfrac{k^2}{k_1} \delta \cos qy\right] & \text{(P)} \end{cases} \tag{6.21}$$

$$P_y \approx \frac{E_0^2}{8\pi} q\delta \sin qy \tag{6.22}$$

In the case of a propagating mode ($q < k$), since k_1 is imaginary we get no transverse modulations for P_x, whereas for evanescent modes ($q > k$) modulations appear. We assume throughout the rest of this derivation $q > k$.

---

[5]This is due to the fact that $\omega_p \gg \omega_0$ for overdense targets. See Sect. 2.3 for more details.

6.1 Rayleigh–Taylor Instability in Radiation Pressure Acceleration

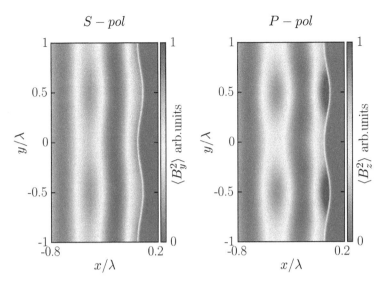

Fig. 6.1 2D simulations performed with *piccante* showing the reflection of a plane wave from a grating target. Simulation were performed with a very low field intensity, so that the dynamics is linear and no deformation of the target surface occurs. The two frames show the temporal average over one laser periodo of B_t^2 (the *squared transverse* component of the magnetic field). Both S-polarization and P-polarization cases are shown. As predicted by the model, when the plane wave is P-polarized the field is locally enhanced in the valleys of the grating. On the contrary, for S-polarized plane waves, the field is mildly enhanced at the grating peaks. Reprinted figure with permission from Andrea Sgattoni, Stefano Sinigardi, Luca Fedeli, Francesco Pegoraro and Andrea Macchi, Physical Review E 91, 013106 [1]. Copyright 2015 by the American Physical Society

Since our goal is to calculate the growth rate of the instability of the surface, we are required to assess if the pressure smooths or enhances surface modulations. Defining P_0 as $P_0 = \dfrac{E_0^2}{4\pi} = \dfrac{2I}{c}$ we get:

$$P_\perp = -\mathbf{P} \cdot \hat{\mathbf{n}} \approx P_0 [1 + K(q)\delta \cos qy] \tag{6.23}$$

where $K(q)$ is defined as ($k_1 = q^2 - k^2$):

$$K(q) = \begin{cases} -\dfrac{\sqrt{q^2-k^2}}{k^2} & \text{(S)} \\ \dfrac{k^2}{k^2-q^2} & \text{(P)} \\ \dfrac{2k^2-q^2}{k^2-q^2} & \text{(C)} \end{cases} \tag{6.24}$$

Equation 6.24 implies that the pressure $P_\perp$ acting on the target surface is modulated differently according to pulse polarization. Since $q > k$, $K(q)$ is positive for P and C polarization and negative for S polarization. As the surface rippling is described by $x_m = (\delta/2) \cos qy$, for P and C polarization the maxima of $P_\perp$ are located in the valleys of the rippling, whereas for S polarization the maxima are located in correspondence of the peaks. Thus irradiating a rippled target with a P- or C-polarized plane wave leads to an enhancement the of surface rippling, while irradiation with an S-polarized plane wave has a smoothing effect on the surface perturbations. Figure 6.1 shows numerical simulations performed with *piccante* which substantially confirm this picture.

We have derived an expression for radiation pressure acting on a shallow rippled surface for various polarizations (Eqs. 6.23 together with 6.24). In the previous derivation, an infinite perfect mirror was considered. We now proceed in our derivation considering a thin foil with a rippled surface. Irradiation of a thin target with an intense EM radiation leads to a forward motion of the whole target. However, since the pressure is applied unevenly on the target surface, the thin foil could be deformed. Here we want to asses the effect of the self-modulated radiation pressure on target dynamics.

For our derivation, we will use Ott's model [25], already exploited in previous works on RPA (see [24]). We restrict our derivation to the non-relativistic case for simplicity. Since we are interested in the first stages of the acceleration process, in which target velocity is presumably still non-relativistic, this assumption is reasonable.

We consider a thin foil of areal density σ, with a pressure $P_\perp$ acting on the left surface (we restrict ourselves to a 2D analysis for simplicity, in agreement with Ott model).

We consider a fluid element with Lagrangian coordinate $\mathbf{r}(\mathbf{r}_0, t)$ of length:

$$d\mathbf{r} = \mathbf{r}\left([x_0 = 0, y_0 + dy_0], t\right) - \mathbf{r}\left([x_0 = 0, y_0], t\right) \quad (6.25)$$

We can now write Newton equation $\mathbf{F} = m\mathbf{a}$ for the fluid element, considering that $m = \sigma dy_0$ and that the pressure $P_\perp$ exerts a force perpendicular to the fluid element surface. The vector tangential to the target surface is given by $\partial_{y_0}\mathbf{r} = \partial_{y_0} x \hat{\mathbf{x}} \partial_{y_0} y \hat{\mathbf{y}}$. Thus, the inward-pointing normal vector to the surface is given by:

$$\hat{\mathbf{n}} = \frac{\hat{\mathbf{x}}\partial_{y_0} y - \hat{\mathbf{y}}\partial_{y_0} x}{\sqrt{(\partial_{y_0} y)^2 + (\partial_{y_0} x)^2}} \quad (6.26)$$

Since the force exerted on the fluid element $\hat{\mathbf{x}}$ is given by

$$\mathbf{F} = P\, \delta s\, \hat{\mathbf{n}} \quad (6.27)$$

where δs is the length[6] of the fluid element and considering that:

$$\delta s = |d\mathbf{r}| = \sqrt{(\partial_{y_0} y)^2 + (\partial_{y_0} x)^2} \quad (6.28)$$

---

[6]We are in 2D geometry.

6.1 Rayleigh–Taylor Instability in Radiation Pressure Acceleration

the equation of motion for the fluid element reads as follows

$$\partial_t^2 \mathbf{r} = \frac{P}{\sigma} \left(\hat{\mathbf{x}} \partial_{y_0} y - \hat{\mathbf{y}} \partial_{y_0} x \right) \tag{6.29}$$

If the pressure P were a constant rather then a function of ξ_x, Eq. 6.29 would have led us to "conventional" RT instability.

Following the Ott model, we can now try to find an approximate solution in the following form. Here we are considering only one mode q at a time; for a more complete solution we should replace second terms in Eq. 6.30 with a summation over q. However, as far as we limit ourselves to the linear phase of the growth of the instability, the surface rippling is simply a linear superposition of the various q modes contributions.

$$\begin{cases} x([0, y_0], t) \sim \xi_0(t) - \xi_x(t) \cos q y_0 \\ y([0, y_0], t) \sim y_0 + \xi_y(t) \sin(q y_0) \end{cases} \tag{6.30}$$

This approximate solution should satisfy Eq. 6.29, with the following boundary conditions:

$$\begin{cases} x(t=0) = 0 \\ y(t=0) = y_0 \\ \frac{\partial x}{\partial t}\big|_{t=0} = 0 \\ \frac{\partial y}{\partial t}\big|_{t=0} = 0 \end{cases} \tag{6.31}$$

This means that at time $t = 0$ we are considering a fluid element at rest located at the target surface. The first condition is satisfied if $\xi_0(0) = 0$, the second condition is automatically satisfied, while the third condition requires that $\partial_t \xi_0|_{t=0} = 0$ (surface at rest at $t = 0$) and that $\partial_t \xi_x|_{t=0} = 0$ (rippling not growing or shrinking at $t = 0$). Similarly the third condition requires $\partial_t \xi_y|_{t=0} = 0$ (rippling not growing or shrinking at $t = 0$).

Using the approximate form for $\mathbf{r}$ in Eq. 6.30 we get:

$$\begin{cases} \partial_t^2 \xi_0(t) - \partial_t^2 \xi_x(t) \cos(q y_0) = \frac{P}{\sigma} \left(1 + \xi_y(t) q \cos q y_0 \right) \\ \partial_t^2 \xi_y(t) = -\frac{P}{\sigma} \xi_x(t) q \end{cases} \tag{6.32}$$

To the lowest order, the first equation of equation system 6.32 reads simply $\partial_t^2 \xi_0(t) = \frac{P}{\sigma}$, which has the very simple solution:

$$\xi_0(t) = \frac{1}{2} \frac{P}{\sigma} t^2 \tag{6.33}$$

In this approximation, the target is not perturbed and translates as a flat slab under the pressure P.

Feeding back Eq. 6.33 into the equation system 6.32 we get:

$$\begin{cases} \partial_t^2 \xi_x(t) = -\dfrac{P}{\sigma}\xi_y(t)q \\ \partial_t^2 \xi_y(t) = -\dfrac{P}{\sigma}\xi_x(t)q \end{cases} \quad (6.34)$$

As noted in [25], Eq. 6.30 does not describe in general a sinusoidal perturbation in Eulerian variables. However, for very small perturbations ($q|\xi_i| \ll 1$) a sinusoidal rippling is a good approximation of the shape of the target surface. Since we are allowed to approximate the deformation of the target surface as a sinusoidal rippling, we may replace the pressure P in Eq. 6.34 using Eq. 6.23: $P \to P_0 [1 + K(q)\xi_x(t)]$.

At first order we get:

$$\begin{cases} \partial_t^2 \xi_x = -\dfrac{P_0[1 + K(q)\xi_x(t)]}{\sigma}\xi_y(t)q \\ \partial_t^2 \xi_y = -\dfrac{P_0[1 + K(q)\xi_x(t)]}{\sigma}\xi_x(t)q \end{cases} \quad (6.35)$$

Looking for exponentially unstable solutions like $\xi_x \sim e^{\gamma t}$ we are led to the following equation for the growth rate:

$$\gamma = \sqrt{\dfrac{P_0}{\sigma}}\left[\sqrt{q^2 + K(q)^2/4} + \dfrac{K(q)}{2}\right]^{1/2} \quad (6.36)$$

The growth rate γ is represented in Fig. 6.2 for the three polarizations S, P, C. As a reference, also the standard result for Rayleigh–Taylor instability $\gamma_{RT} = \sqrt{P_0 q/\sigma}$ is reported. The growth rate γ_{RT} is still valid for $q < k$ (when the periodicity of the rippling is smaller than laser wavelength and hence no effect of the radiation pressure modulation is present). A careful discussion of the Laser-driven RTI growth rate is required.

We have derived the growth rate of the instability for the three polarizations S, P and C. However, laser-plasma interaction in the high intensity regime is strongly dependent on pulse polarization. Indeed, for linearly polarized laser pulses at normal incidence the surface rippling may be "washed out" by the quiver motion of the electrons, even at moderate a_0. As a consequence, we expect the above theory to be appropriate for C polarization, whereas discrepancies with simulations and experiments are foreseen for linear polarization. However, it is worth to remark that the preferred option for radiation pressure acceleration experiment is irradiation with circular polarization, because it preserves the integrity of the thin target for a longer time.

Moreover, as already mentioned, we expect the conventional γ_{RTI} to be valid for $q/k < 1$ (modes with periodicity longer that one wavelength) rather that the laser

6.1 Rayleigh–Taylor Instability in Radiation Pressure Acceleration

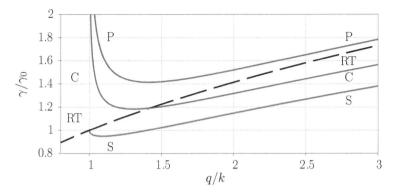

Fig. 6.2 The figure illustrates radiation pressure RTI growth rate of a thin foil irradiated with a S-, P-, C- polarized laser (*solid curves*). The *dashed curve* is the "standard" RTI growth rate for a flat thin foil: $\gamma_{RTI} = \sqrt{P_0 q/\sigma}$. For $q/k < 1$ k_1 is imaginary and the other curves are reduced to γ_{RTI}. In the graph γ is normalized over $\gamma_0 = \sqrt{P_0 k/\sigma}$. Reprinted figure with permission from Andrea Sgattoni, Stefano Sinigardi, Luca Fedeli, Francesco Pegoraro and Andrea Macchi, Physical Review E 91, 013106 [1]. Copyright 2015 by the American Physical Society

driven RTI growth rate. Finally, further considerations suggest that instability modes with $q/k > 2$ should be strongly suppressed (in analogy with waveguides, if the peak-to-peak size of the rippling is smaller than $\lambda/2$, light cannot penetrate in the valleys, thus preventing a further development of the instability).

In conclusion, we expect our theory to be a faithful description of the onset of the instability for C-polarized laser pulses. The predicted growth rate as a function of q/k is given by the blue curve in Fig. 6.2 between $q/k = 1$ and $q/k = 2$ and by the dashed curve for $q/k < 1$, whereas for $q/k > 1$ the growth rate should be strongly suppressed. Since the blue curve is divergent for $q/k \to 1^+$, we expect the dominant scale-length of the surface rippling to be close to the laser wavelength.

It is important to point out that we've developed a 2D model of laser-driven Rayleigh–Taylor instability, ignoring the dynamics in the third direction. For this reason the growth rate of the instability are different for P and S polarization. Obviously, for normal incidence, there should be no difference between these polarizations (except for a 90° rotation of the system). The results we've obtained for linear polarization simply suggest that the development of the instability would lead to asymmetric structures.

6.1.3 Numerical Simulations

Numerical simulations for this work were performed partially with *piccante* code (2D simulations) and partially with ALaDYn [26] code (3D simulations). The exact source code and settings used for 2D simulations with *piccante* can be retrieved at [27].

In order to test a regime of "pure" RPA the numerical simulations were performed with intensities significantly greater then what can be achieved by modern laser systems ($a_0 = 66$ for the 2D simulations and $a_0 = 280$ for the 3D simulations). However these intensity regimes are foreseen for future generation laser systems.

2D Simulations

2D simulations were performed with the aim of testing the analytical model described in Sect. 6.1.2. Performing simulations in 2D allows for a high spatial resolution, which would be extremely demanding for a 3D simulation.

The main predictions of the theoretical model for circularly polarized light is a strong resonance if the instability scalelength δ is close to the laser wavelength λ. Plasmonic enhancement of the local field occurs at this resonance.

In order to test these predictions, we performed a numerical simulation campaign on FERMI supercomputing machine. The results shown in Fig. 6.3 are referred to

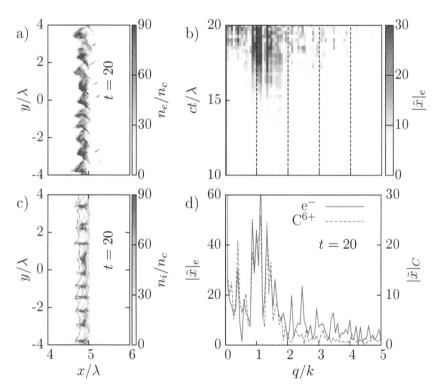

Fig. 6.3 Analysis of the transverse modes in 2D plane wave simulations. (**a**), (**c**) Charge density ($t = 20\lambda/c$) of electrons and ions. For each time step the longitudinal position $x = x(y, t)$ of the vacuum-plasma interface was reconstructed as a function of the transverse coordinate y. **b** Temporal evolution of the Fourier transform $\tilde{x}(q, t)$ for electrons. **d** Comparison of $\tilde{x}$ for electrons (*thick line*) and carbon ions (*dashed line*) at t = 20. Reprinted figure with permission from Andrea Sgattoni, Stefano Sinigardi, Luca Fedeli, Francesco Pegoraro and Andrea Macchi, Physical Review E 91, 013106 [1]. Copyright 2015 by the American Physical Society

6.1 Rayleigh–Taylor Instability in Radiation Pressure Acceleration

Table 6.1 Parameter list for 2D simulations

Simulation parameters	2D simulations
Simulation box	$45\lambda \times 15\lambda$
Resolution (points per λ)	204
Target density	$37\,n_c$
Target thickness	$0.58\,\lambda$
Particles per cell (electrons)	81
Composition	C ($Z/A = 0.5$)
Contaminants	No
Laser polarization	C, P
Laser a_0	66
Laser type	$\cos^2$ plane wave
Laser duration FWHM	$12.4\,\lambda/c$

a simulation where the laser intensity $a_0 = \sqrt{\dfrac{I}{2m_e c^3 n_c}}$ was 66, the target was thin ($d = 0.58\lambda$) and overdense $n_e = 37n_c$.[7] As usual for solid target simulations, a high spatial resolution was required ($\Delta x = \Delta y = \lambda/204$). Simulation parameters are detailed in Table 6.1.

Results shown in Fig. 6.3 substantially confirm the theoretical considerations, since the dominant modes of the instability are concentrated in the $[k, 2k]$ range ($k = 2\pi/\lambda$).

As expected, for linear P polarization, the laser-driven RTI is quenched: intense electron heating processes quickly wash out the surface rippling.

3D Simulations

3D numerical simulations were performed with ALaDyn Particle-In-Cell code [26],[8] with parameters close to those of the RPA scenario studied in [28].

The simulation box is $93\,\lambda$ wide along $\hat{\mathbf{x}}$ (the laser propagation direction) and $120\,\lambda$ wide along $\hat{\mathbf{y}}$ and $\hat{\mathbf{z}}$ (the transversal directions). In order to save computational resources, a simulation grid stretched along $\hat{\mathbf{y}}$ and $\hat{\mathbf{z}}$ is adopted. This allows to save grid nodes along those directions while keeping at the same time a high resolution in the interaction region, where it is crucial to resolve laser-matter interaction. Indeed the resolution in the central region ($93\lambda \times 60\lambda \times 60\lambda$) is 44 points per λ along $\hat{\mathbf{x}}$ and 22 points per λ along the transverse directions, while the total grid size is only $4096 \times 1792 \times 1792$ grid cells. A large grid along $\hat{\mathbf{y}}$ and $\hat{\mathbf{z}}$ is mainly needed to prevent electrons from reaching the borders of the box.[9]

---

[7] Not far from the density of completely ionized solid hydrogen.

[8] When the simulation was performed, *piccante* code wasn't already available.

[9] To the knowledge of the author, there isn't a completely "clean" way to deal with charged particles reaching the borders of the simulation box. An exception is represented by particles reaching

Table 6.2 Parameter list for 3D simulations

Simulation parameters	3D simulations
Simulation box	$93\lambda \times 60\lambda \times 60\lambda$
Resolution (points per λ)	[44, 22, 22] (central region)
Target density	$64\,n_c$
Target thickness	$1\,\lambda$
ParticlesPerCell (electrons)	64
Composition	C ($Z/A = 0.5$)
Contaminants	Yes
Laser polarization	C, P
Laser a_0	198 (C), 280 (P)
Laser type	Gaussian
Laser waist	$6\,\lambda$
Laser duration FWHM	$9\,\lambda/c$

The target is composed by a thin, $1\,\lambda$ thick, C^{6+} foil ($Z/A = 0.5$) and by a $\lambda/22$ thick hydrogen contaminant layer on the back side of the carbon foil. Target density is $64 n_c$ for the foil and $8 n_c$ for the contaminant layers. All the species (electrons, carbon ions and hydrogen ions) are sampled with 64 particles per cell, in order to resolve the charge density down to a fraction of the critical density n_c. The total number of particles is $\approx 2 \times 10^{10}$.

As far as the laser pulse is of concern, the transverse profile is Gaussian, with a waist diameter $w = 6\lambda$, while the longitudinal profile is $\cos^2$-like, with a FWHM duration of $\tau_p = 9\lambda/c$ (all the aforementioned quantities are referred to the fields, not to the pulse intensity). Both linear and circular polarization were tested and will be referred in the following, respectively, as LP and CP. For CP a laser amplitude $a_0 = 198$ was used, while for LP the laser amplitude was $198\sqrt{2}$.

The simulations were run on 16384 cores at FERMI BG/Q supercomputer (CINECA, Bologna, Italy). The parameters used for 3D simulations are summed up in Table 6.2.

In Fig. 6.4 a snapshot at intermediate times of electron and ion density is shown for the CP case. The development of net-like structures of the ion density is evident from the figure. For LP the ion density develops striped structures stretched along the polarization direction. In order to highlight the differences between the CP case and the LP case, excluding at the same time effects related to the laser pulse shape, we performed also 3D plane wave simulations. These simulations were performed with *piccante* particle-in-cell code. Apart from the plane wave pulse, the main difference with Gaussian pulse simulations is a significantly reduced transverse size of the simulation. Since the target is irradiated with a plane wave, the system

---

(Footnote 9 continued)
a moving border, like when *moving window* method is used, provided that the border moves at the speed of light.

6.1 Rayleigh–Taylor Instability in Radiation Pressure Acceleration

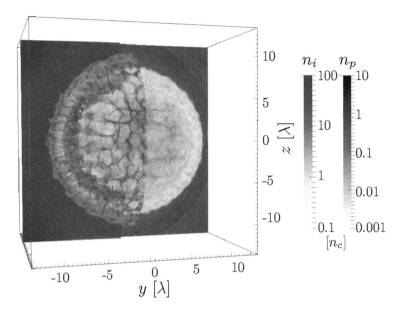

Fig. 6.4 The picture shows a snapshot (t = 30 T) of the 3D RPA simulation, depicting the density of both proton (*dark green tones*) and carbon (*light blue tones*) densities. In order to make the carbon ion density visible, the proton density is shown only on the left part ($y < 0$) of the image. Reprinted figure with permission from Andrea Sgattoni, Stefano Sinigardi, Luca Fedeli, Francesco Pegoraro and Andrea Macchi, Physical Review E 91, 013106 [1]. Copyright 2015 by the American Physical Society (colour figure online)

is invariant for translations along the transverse direction (periodic boundary conditions are applied). Thus, it is sufficient to have a simulation box large enough to allow the development of the structures. A transverse size of 5λ along $\hat{\mathbf{y}}$ and $\hat{\mathbf{z}}$ was chosen for plane wave simulations. Due to the limited computational resources required by these simulations, the grid was uniform (i.e. not stretched).

Figure 6.5 shows two snapshots of the density of carbon ions for CP (left panel) and LP (right panel), taken at the same simulation time. Apart from the pulse polarization, the other parameters were identical.

In the CP case the ion density showed a pattern of hexagonal-like denser structures, with "holes" in between which were significantly depleted of particles. Almost a factor of two is observed between the density of the structures and the density of the holes. The self-organization of the density in hexagonal structures corresponds closely to a theoretical prediction, based on symmetry arguments, for a stable structure of the flow in the nonlinear 3D development of the RTI [1, 29, 30]. It is worth to mention that the structures formed in the CP case are an example of spontaneous symmetry breaking occurring in a classical system [32]. Indeed, the system in its initial state is invariant for continuous rotations along $\hat{\mathbf{x}}$ and for continuous translations along a transverse direction. The development of laser-driven RTI reduces the symmetry of the system to the discrete "wall-paper" group p6mm [33]. In mathematics,

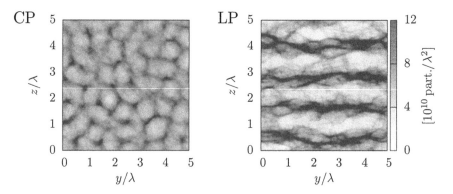

Fig. 6.5 The figure shows the areal density of carbon ions at t = 15 T in 3D simulations with the same parameters as in Fig. 6.4, but for an incident plane wave, for circular (CP) and linear (LP) polarization. Reprinted figure with permission from Andrea Sgattoni, Stefano Sinigardi, Luca Fedeli, Francesco Pegoraro and Andrea Macchi, Physical Review E 91, 013106 [1]. Copyright 2015 by the American Physical Society

the symmetry properties of repetitive patterns are classified according to "wall-paper" groups (the name is due to the fact that wall and floor tessellations based on repetitive patterns are often used in architecture). Wall-paper groups are distinguished by number of discrete rotation centres and their order, by symmetry for discrete translations and by symmetry for flipping. The order of a rotation is defined by the number of discrete rotations to perform before recovering the full 360° angle (e.g. a rotation centre of order six means that the pattern is symmetry for 60° rotations). In particular, p6mm group is characterized by one rotation centre of order six, two rotation centres of order three and three rotation centres of order two. The two rotation centres of order three differ by a rotation of 60° (or 180°), while the three rotation centres of order two differ by a rotation of 60° Additionally, a p6mm is symmetric for reflection in six distinct directions. It has also reflections in six distinct directions. There are additional symmetries for glade reflections[10] in six distinct directions. Figure 6.6 shows an example of a mosaic pattern with p6mm symmetry group.

Also for LP case, laser interaction with the targets results in the formation of structures in the ion density. These structures are strongly elongated along the polarization direction (in this case there is no spontaneous symmetry breaking, because the linear polarization sets a preferential transverse direction). The elongation of the structures confirms that the intense electron heating mechanisms wash out the target surface modulations, preventing the growth of structures along the electric field direction.

The data was shown for ions, however similar structures develop also for the electron density. An apparent faster growth of the structures is observed for LP, if compared to CP.

---

[10]Reflections associated with a translation.

Fig. 6.6 The figure is an example of a Byzantine mosaic in Rome with p6mm symmetry. Picture reproduced from [31]

6.1.4 Conclusions

We showed that self-consistent modulation of radiation pressure and plasmonic enhancement at a rippled surface strongly affect the laser-driven Rayleigh–Taylor instability, setting a dominant scale close to the laser wavelength as observed in simulations. We calculated the growth rate of laser-driver RTI for different laser polarizations. Three-dimensional CP simulation show the formation of netlike structures with approximate hexagonal wallpaper symmetry, in agreement with theoretical predictions. LP simulations are affected by strong electron heating processes and structures along polarization directions are washed out.

The laser-driven RTI rises some concerns over the light sail ion acceleration scheme. Indeed, the development of surface rippling may lead to an early onset of foil transparency, with a consequent breakout of the target and efficiency losses. However, in 3D PIC simulations of the light sail regime, the early growth of RTI doest not prevent the ions from reaching high energies (see [19, 34]). Even though not preventing ion acceleration, laser-driven RTI determines a strong modulation of the accelerated beam (strong modulation of laser-accelerated ion beams were observed in [2]).

Field modulation and local enhancement were studied in the context of radiation pressure acceleration. However, they may play a role in other scenarios involving laser interaction with overdense plasmas, like generation of current filaments from the interaction surface (which in several simulations is correlated with a local rippling see [35, 36]). Indeed, a transverse modulation of the EM field translates into a modulation of the $\mathbf{v} \times \mathbf{B}$ force, which in turn leads to a modulation of the energy of the accelerated electrons. These accelerated electrons may provide a seed for the filamentation instability (see [37]) with a spatial periodicity equal to the laser wavelength λ_L, explaining why λ_L is the preferred scale for the filaments (see [38, 39]).

6.2 Plasmonic Effects in High Order Harmonic Generation from Grating Targets

This section deals with plasmonic effects in HHG from grating targets irradiated with ultra-intense laser pulses. The generation of High Order Harmonics and attosecond pulses has led in the recent past to very interesting results. Standard high-order harmonics sources are based on gaseous targets: the ionization of the atoms and the subsequent re-collission of the extracted electron with its parent ion is at the basis of the coherent harmonic emission process [40, 41]. Since HHG in this scheme is suppressed beyond a certain value of the laser intensity, there is a limit to the maximum attainable intensity of the harmonic source. HHG with solid targets could represent an interesting solution, since no fundamental limit to the intensity of the laser pulse exists. On the contrary, HHG should be enhanced for higher pulse intensities. However, technical difficulties have hampered so far the use of HHG from solid targets (see Sect. 6.2.1).

Solid grating targets are attractive for HHG, since the diffraction should separate angularly the harmonics, easing the extraction of a selected family of harmonics (this scheme has been studied in the past). Here the idea is to study HHG with a grating target irradiated close to the angle expected for SW excitation. The idea is that the plasmonic field enhancement which characterizes the resonance condition should in turn enhance HHG.

6.2.1 Introduction on HHG with Laser-Based Sources

High-order Harmonic Generation via laser interaction with gaseous targets (see [40, 42]) is a very active research field. Indeed, it allows to generate extremely-short *attosecond* pulses (see [43]), which can be exploited for a wide range of applications (e.g. imaging and control of electron motion in atoms and molecules, see [44]).

6.2 Plasmonic Effects in High Order Harmonic Generation from Grating Targets

A detailed overview of this vibrant research field is beyond the scope of this document and the interested reader is referred to [45].

HHG from Solid Targets

The generation of High-order harmonics in intense fs laser interaction with solid density plasma was first observed in [46]. Harmonic generation is usually observed in the presence of a steep density gradient of the plasma. As pointed out in [47], two main physical processes contribute to HHG with solid targets: *Coherent Wake Emission* (CWE) [48] and *Relativistically Oscillating Mirror* (ROM) [49–52]. The first mechanism is active already at quite low, non relativistic, laser intensities ($I \sim 10^{14}$ W/cm^2) and is able to generate harmonics up to the plasma frequency ω_p. CWE involves electron bunches extracted from the plasma and subsequently re-injected in a fraction of the optical cycle. The ROM mechanism instead is based on the non-linear effect of the relativistic oscillation of the vacuum-plasma boundary, which, being overdense, acts as a mirror. The theory of ROM is complicated, but a simple model is provided hereunder (see [11] for details). We consider a steep plasma density with oscillating boundary:

$$n_e = n_0 \Theta(x - \xi(t)) \tag{6.37}$$

where

$$\xi(t) = \xi_s \sin(\omega_0 t) \tag{6.38}$$

is the position of the vacuum-plasma interface.

Suppose that there is an incoming laser pulse $E_L \sin(\omega t)$ and that we observe the reflected radiation at a fixed position $-d$ in front of the mirror. The time needed for a photon to travel from the observation point to the mirror, get reflected and reach again the observation point is given by $t_{ret} = t + 2(d + \xi_s \sin(\omega_0 t))$. Thus we obtain the following expression for a reflected laser pulse:

$$E_r = E_L \sin(\omega t_{ret}) = E_L \sin(\omega t - \frac{\omega}{c} \xi_s \sin(\omega_0 t) + \phi_0) \tag{6.39}$$

where ϕ_0 is a phase factor. Equation 6.39 clearly contains a non-linear term. Figure 6.7 shows how a reflected pulse is affected by the moving mirror. The Fourier transform of the reflected pulse shows clearly a high harmonic content. The model provided here is over-simplified, but it helps to understand the main features of the physical process at play. A proper derivation of the ROM mechanism is provided by [51, 52]. Unlike CWE, ROM process requires relativistic laser intensities. In the strong relativistic case ($a_0 >> 1$), the intensity of the harmonic orders is expected to scale as $n^{-8/3}$ (n is the harmonic order), up to a maximum harmonic order n_{max} which is proportional to γ^3 (where γ is calculated with the peak velocity of the moving mirror).

The laser wavelength of a Ti:Sapphire system is $\lambda \sim 800$ nm. Harmonics from $n = 7$ up to $n = 80$ are within the XUV *eXtreme Ultra-Violet* region of the EM spectrum ($\lambda_{XUV} = 10 - 124$nm), which may have applications for XUV diffraction experiments [53] or for the study of XUV ionization of molecules [54] (as noted in [47]). Since the aforementioned processes can be extended up to relativistic

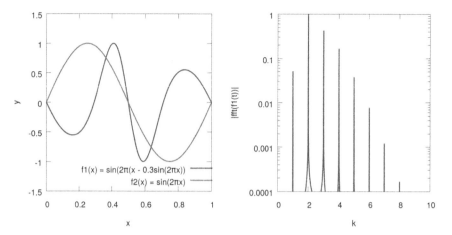

Fig. 6.7 The *panel* on the left shows how a relativistically oscillating mirror reflects a simple sin electromagnetic wave. The *green curve* is the incoming wave, while the *violet curve* is the reflected pulse. The steepening of the reflected pulse suggests an abundant harmonic content in the reflected pulse, as confirmed by the Fourier transform reported in the *panel* on the right (colour figure online)

intensities, HHG with solids is attractive for the development of intense, pulsed XUV sources. Unlike HHG from gaseous targets, harmonic emission would last for the whole laser-target interaction time (tens of femtoseconds). This means that, unless gating techniques are used, HHG from solid targets cannot be used for the generation of attosecond pulses [55, 56].

6.2.2 Grating Targets as a HHG Source

The study of grating targets as a HHG source has been pursued in the recent past, bot numerically and experimentally (see [6, 47, 57–59]). The main advantage of a grating over a flat mirror is the possibility to disperse angularly the harmonics, allowing to select a *quasi-monochromatic* XUV beam. Indeed, a beam selected in a given direction would contain a harmonic order and its multiples.

The expected angular distribution for the harmonics generated with a solid density target is given by [6]:

$$\frac{n\lambda}{m} = d|\sin(\theta_{inc}) - \cos(\alpha)| \qquad (6.40)$$

where θ_{inc} is the angle of incidence, n is the interference order, m is the harmonic order and d is the grating spacing.

6.2 Plasmonic Effects in High Order Harmonic Generation from Grating Targets

Table 6.3 Parameter list for 2D HHG with gratings simulations

Simulation parameters	2D simulations
Simulation box	$80\lambda \times 80\lambda$
Resolution (points per λ)	[102.4, 102.4]
Target	
Density	$64.0\, n_c$
Thickness	$1.0\, \lambda$
Peak-to-valley depth	$0.25\, \lambda$
Periodicity	$1.52\, \lambda$ (G20), $1.35\, \lambda$ (G15), ∞ (F)
Particles per cell (electrons)	64
Particles per cell (ions)	16
Laser	
Polarization	P
a_0	5
Waist	$5\, \lambda$
Type	Gaussian
Pulse incidence	$0° - 30°$ (various)
Duration FWHM	$12\, \lambda/c$

Plasmonic Effects on HHG with Gratings

If a grating is used as a target for HHG, not only dispersion, but also plasmonic effects may play a role. Indeed, as discussed in Sect. 6.1, when a grating is irradiated, significant field enhancement effects may take place. The field intensity affects the efficiency of HHG and thus plasmonic effects should be able to influence the intensity of the emitted harmonics.

In order to study the aforementioned process, an extensive parametric scan was completed with 2D simulations (performed with *piccante*). Table 6.3 lists the parameters used for this numerical investigation campaign. The resolution should be relatively high, in order to correctly resolve the higher order harmonics. As a rule of thumb, an EM wave with wavelength λ is correctly reproduced in a PIC code only if the resolution is at least $\lambda/10$. This means that, with a resolution of ~ 100 points per λ_l (the wavelength of the laser), up to the tenth harmonic order can be resolved reliably. Depending on the simulation, the target was a grating with periodicity 1.52λ (G20, resonance expected at $20°$), a grating with periodicity 1.35λ (G15, resonance expected at $15°$) and a flat target (F, no modulation of the irradiated surface). The thickness of the target was 1λ and the peak-to-valley depth of the modulations was $\lambda/4$. The density of the target was chosen to be $n_e/n_c = 64$ (strongly over-dense target, though not as dense as real solid targets). 64 particles per cell were used for electrons, in order to resolve the density down to $\sim 1\, n_c$. Instead, since ions are expected to play a negligible role in the process, their density was sampled with only 16 particles per cell. The simulated laser was a gaussian pulse. P-polarized, with

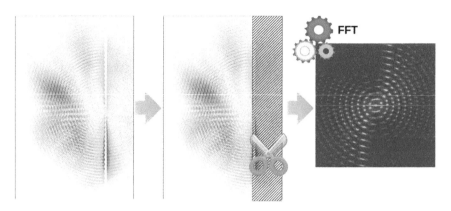

Fig. 6.8 Simulation data are processed as shown in the figure. A snapshot of the B_z component of the magnetic field is taken at the end of the interaction of the laser pulse with the target. Only the output region with $x < 0$ ($x = 0$ is the target surface) is considered. A fourier transform is then performed with *GNU Octave*

$a_0 = 5$ (fully relativistic), a waist equal to 5λ and a FWHM duration of $12\lambda/c$. The angle of incidence of the pulse was varied between $0°$ and $30°$.

The procedure outlined in Fig. 6.8 is used to analyse the simulation results. This procedure is applied to the $\hat{z}$ component of the magnetic field (the only component of the magnetic field for P-polarized pulses in a 2D simulation). The simulation is run long enough to observe the end of the laser-target interaction. Then, the right half of the simulation box is cut out (the exact left limit of the cut is fixed at the average position of the target surface[11]). Finally, a 2D Fourier transform is performed to obtain $\tilde{B}_z(k_x, k_y)$ (where k_x and k_y are the wave-vector components, respectively along $\hat{x}$ and $\hat{y}$). Figure 6.9 shows a sequence of graphs of $\tilde{B}_z(k_x, k_y)$ for G20 and various angles of incidence. The figure suggests that an enhanced harmonic generation takes place for angles of incidence between $20°$ and $25°$, so for angles of incidence slightly larger than the expected resonance at $20°$. The angular distribution of a given harmonic order can be shown plotting $\tilde{B}_z(k_x, k_y)$ over a circular path of radius $k = n$ (where n is the desired harmonic order). Figure 6.10 shows the angular distribution of the harmonics emitted by a G20 and an F target. The 8th and the 9th harmonic are shown. For F the harmonics are emitted essentially along the specular reflection angle and the efficiency grows together with the angle of incidence. For G20 the graphs are significantly more complicated, since for each angle of incidence harmonics are emitted in several directions. However, particularly strong signals are emitted close to the target tangent (between $75°$ and $90°$) for angles of incidence slightly larger than the expected resonance ($20°$–$25°$). Since the scale on the $\hat{y}$ axis is the same for all the graphs the curves can be compared directly: the efficiency for harmonic generation seems to be significantly higher for G20 than for F, especially for the 9th harmonic. In

---

[11] i.e. for a grating target if x_p is the x coordinate of the peaks and x_v is the x coordinate of the valleys, x_{cut} is fixed at $(x_p + x_v)/2$.

6.2 Plasmonic Effects in High Order Harmonic Generation from Grating Targets

Fig. 6.9 $\tilde{B}_z(k_x, k_y)$ for a G20 target irradiated at several angles of incidence. Enhanced harmonic generation is evident for angles of incidence between 20° and 25°

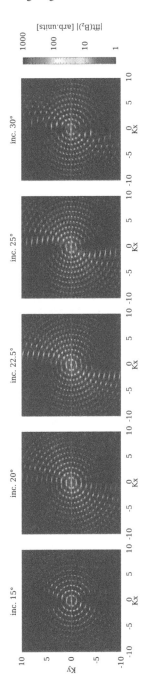

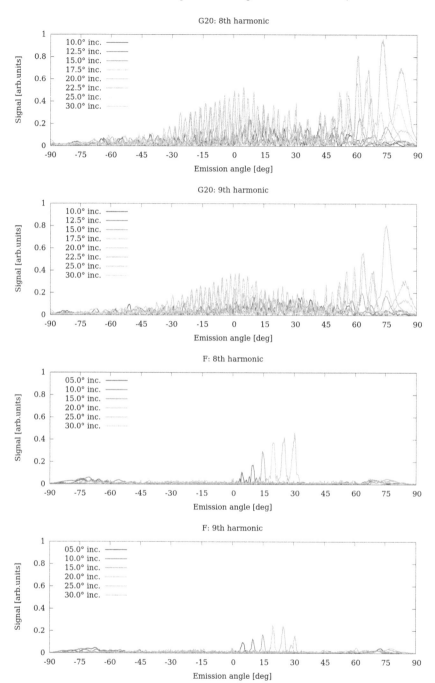

Fig. 6.10 Angular distribution of the 8th and 9th harmonic emitted by a G20 and an F target at various angles of incidence. The same scale is adopted for the axes, so the signals can be compared directly. The angles span all the front side of the target. The laser comes with a negative angle of incidence, 0° is the target normal while ±90° are the target tangent

6.2 Plasmonic Effects in High Order Harmonic Generation from Grating Targets

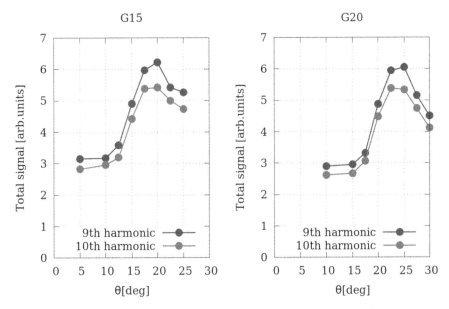

Fig. 6.11 These graphs have been obtained integrating the absolute value of $\tilde{B}_z(k_x, k_y)$ in the annular region $8.5 < k < 9.5$ for the 9th harmonic and in the annular region $9.5 < k < 10.5$ for the 10th harmonic. The *left panel* shows data for G15, while the *panel* on the right shows data for G20. The value of the integral in the annular region is reported as a function of the angle of incidence

order to compare the efficiency of harmonic conversion, an integration of $\tilde{B}_z(k_x, k_y)$ in an annular region around the circle $k = n$ can be performed. Figure 6.11 shows the result for G15 and G20 irradiated at various angles of incidence. The value of the integral reaches a peak around 17.5°–20° for G15 and around 22.5°–25° for G20. After having reached the peak, the integral decreases for larger angles of incidence. The value reached at the peak is essentially the double of that of the smaller angles of incidence.

6.2.3 Conclusions

Although preliminary, the results shown in this section suggest that plasmonic effects could be relevant for HHG with solid grating targets. In particular, irradiating a grating target with an angle slightly larger than the angle expected for SW excitation results in a significant enhancement of HHG in certain directions. The increased efficiency in high order harmonic conversion is presumably due to plasmonic field enhancement. The study of this effect, which could help in the realization of very intense quasi-monochromatic XUV sources, will be pursued in the future.

6.3 Energy Concentration Schemes?

This section is intended to present some preliminary work on energy concentration schemes in the high field regime. The aim is to discuss at what extent schemes from traditional plasmonics can be ported in the high field regime and to individuate possible routes to purse energy concentration in these conditions. Achieving field enhancement in ultra intense laser-matter interaction could lead to field intensities beyond what can be obtained with present and foreseen laser facilities.

Several schemes for extreme field enhancement exist in traditional plasmonics (see [8] and Sect. 2.3.2): tapered waveguides, tapered tips, kissing cylinders[12]... Of course schemes based on the use of dielectrics cannot be ported in high field regime (unless the dielectric can be replaced with the vacuum). Indeed, at high laser intensities any material becomes a plasma (and thus a very good conductor) within a single laser cycle. Two schemes could be an interesting starting point for the study of this phenomenon: tapered metallic tips in the vacuum and tapered waveguides. In both cases the coupling of the EM wave with the laser pulse should be obtained with a grating.

The scheme presented here mimics the nanotip concentrator of traditional plasmonics [8, 61]: a SP is induced with laser-grating interaction and is made to propagate along a tapered wedge. The idea is that energy concentration should be observed at the tip.

The parameters of the simulation are reported in Table 6.4. As Fig. 6.12 shows, energy concentration is actually observed at the final portion of the tapered tip.

Table 6.4 Parameter list for 2D simulation of *plasmonic* tapered tip

Simulation parameters	2D simulation
Simulation box	$35\lambda \times 32.5\lambda$
Resolution (points per λ)	[200, 220.6]
Target density	$64\, n_c$
Target thickness	$1.0\, \lambda$
Grating periodicity	$2\, \lambda$ (resonance at 30°)
Grating peak-to-valley depth	$0.3\, \lambda$
Tapering angle	11.3°
Particles per cell (electrons)	64
Composition	C (Z/A = 0.5)
Laser polarization	P
Laser a_0	1
Laser waist	5λ
Laser duration FWHM	$9\lambda/c$
Laser angle of incidence	30°

---

[12] A plasmonic scheme consisting in two juxtaposed cylinders [60].

6.3 Energy Concentration Schemes?

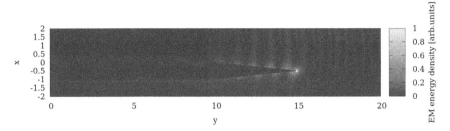

Fig. 6.12 Example of a tapered tip

Besides the study of tapered tips, also the study of tapered plasmonic waveguides may be an interesting route to pursue plasmonic energy concentration in the high field regime. Figure 6.13 is provided to give a sketch of a possible concentrating schemes: grated walls are used to couple the laser to surface plasmons, which should then propagate inside a tapered structure.

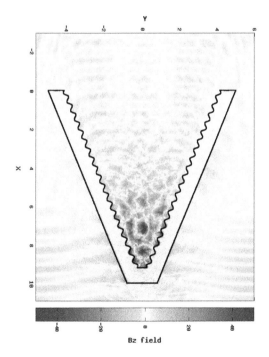

Fig. 6.13 Example of a tapered waveguide

6.3.1 Conclusions

The results shown in this section should be considered as preliminary. Several issues should be addressed to pursue the development of plasmonic energy concentration schemes in the high field regime. In particular, a delicate issue is related to the survival of the sub-wavelength structures for long enough to allow for the plasmonic effects to take place. For example, a sub-wavelength tapered waveguide may be filled with plasma extracted from its wall, preventing the pulse to reach the tip. Similarly, the tip of a tapered tip concentrator may expand due to the intense EM fields.

In conclusion, the topic of plasmonic energy concentration schemes in the high field regime is very attractive. However, further investigation is needed to assess its feasibility.

References

1. A. Sgattoni, S. Sinigardi, L. Fedeli, F. Pegoraro, A. Macchi, Laser-driven Rayleigh–Taylor instability: plasmonic effects and three-dimensional structures. Phys. Rev. E **91**, 013106 (2015)
2. C.A.J. Palmer, J. Schreiber, S.R. Nagel, N.P. Dover, C. Bellei, F.N. Beg, S. Bott, R.J. Clarke, A.E. Dangor, S.M. Hassan, P. Hilz, D. Jung, S. Kneip, S.P.D. Mangles, K.L. Lancaster, A. Rehman, A.P.L. Robinson, C. Spindloe, J. Szerypo, M. Tatarakis, M. Yeung, M. Zepf, Z. Najmudin, Rayleigh–Taylor instability of an ultrathin foil accelerated by the radiation pressure of an intense laser. Phys. Rev. Lett. **108**, 225002 (2012)
3. H. Daido, M. Nishiuchi, A.S. Pirozhkov, Review of laser-driven ion sources and their applications. Rep. Prog. Phys. **75**(5), 056401 (2012)
4. A. Macchi, M. Borghesi, M. Passoni, Ion acceleration by superintense laser-plasma interaction. Rev. Mod. Phys. **85**, 751–793 (2013)
5. U. Teubner, P. Gibbon, High-order harmonics from laser-irradiated plasma surfaces. Rev. Mod. Phys. **81**, 445–479 (2009)
6. M. Cerchez, A.L. Giesecke, C. Peth, M. Toncian, B. Albertazzi, J. Fuchs, O. Willi, T. Toncian, Generation of laser-driven higher harmonics from grating targets. Phys. Rev. Lett. **110**, 065003 (2013)
7. M.I. Stockman, Nanofocusing of optical energy in tapered plasmonic waveguides. Phys. Rev. Lett. **93**, 137404 (2004)
8. D.K. Gramotnev, S.I. Bozhevolnyi, Plasmonics beyond the diffraction limit. Nat. Photonics **4**, 83–91 (2010)
9. D.H. Sharp, An overview of Rayleigh–Taylor instability. Phys. D **12**(1–3), 3–18 (1984)
10. B. Eliasson, Instability of a thin conducting foil accelerated by a finite wavelength intense laser. New J. Phys. **17**(3), 033026 (2015)
11. A. Macchi, *A Superintense Laser-Plasma Interaction Theory Primer* (Springer, Dordrecht, 2013)
12. A. Macchi, S. Veghini, F. Pegoraro, "Light sail" acceleration reexamined. Phys. Rev. Lett. **103**, 085003 (2009)
13. J.N. Bahcall, A.M. Serenelli, S. Basu, New solar opacities, abundances, helioseismology, and neutrino fluxes. Astrophys. J. Lett. **621**(1), L85 (2005)
14. F. Brunel, Anomalous absorption of high intensity subpicosecond laser pulses. Phys. Fluids (1958-1988) **31**(9), 2714–2719 (1988)

15. P. Gibbon, Efficient production of fast electrons from femtosecond laser interaction with solid targets. Phys. Rev. Lett. **73**, 664–667 (1994)
16. S.S. Bulanov, A. Brantov, V.Y. Bychenkov, V. Chvykov, G. Kalinchenko, T. Matsuoka, P. Rousseau, S. Reed, V. Yanovsky, D.W. Litzenberg, K. Krushelnick, A. Maksimchuk, Accelerating monoenergetic protons from ultrathin foils by flat-top laser pulses in the directed-Coulomb-explosion regime. Phys. Rev. E **78**, 026412 (2008)
17. A. Macchi, S. Veghini, T.V. Liseykina, F. Pegoraro, Radiation pressure acceleration of ultrathin foils. New J. Phys. **12**(4), 045013 (2010)
18. G. Marx, Interstellar vehicle propelled by terrestrial laser beam. Nature **211**, 22–23 (1966)
19. A. Sgattoni, S. Sinigardi, A. Macchi, High energy gain in three-dimensional simulations of light sail acceleration. Appl. Phys. Lett. **105**(8), 084105 (2014)
20. M. Tamburini, F. Pegoraro, A. Di Piazza, C.H. Keitel, A. Macchi, Radiation reaction effects on radiation pressure acceleration. New J. Phys. **12**(12), 123005 (2010)
21. R. Capdessus, P. McKenna, Influence of radiation reaction force on ultraintense laser-driven ion acceleration. Phys. Rev. E **91**, 053105 (2015)
22. D. Doria. Ion acceleration from ultrathin foils: dependence on target thickness and laser polarization, in *Oral Contribution at the 42nd EPS Conference on Plasma Physics* (Lisbon, 22–26 June 2015)
23. S. Goede, C. Roedel, M. Gauthier, W. Schumaker, M. MacDonald, J. Kim, R. Mishra, F. Fiuza, S. Glenzer, K. Zeil et al., Effect of the Rayleigh–Taylor-instability on radiation-pressure-accelerated protons from solid-density hydrogen jets. Bull. Am. Phys. Soc. **60** (2015)
24. F. Pegoraro, S.V. Bulanov, Photon bubbles and ion acceleration in a plasma dominated by the radiation pressure of an electromagnetic pulse. Phys. Rev. Lett. **99**, 065002 (2007)
25. E. Ott, Nonlinear evolution of the Rayleigh–Taylor instability of a thin layer. Phys. Rev. Lett. **29**, 1429–1432 (1972)
26. C. Benedetti, A. Sgattoni, G. Turchetti, P. Londrillo, ALaDyn: a high-accuracy pic code for the Maxwell–Vlasov equations. IEEE Trans. Plasma Sci. **36**(4), 1790–1798 (2008)
27. A. Sgattoni, S. Sinigardi, L. Fedeli, piccante, release tailored for RT paper (2014)
28. M. Tamburini, T.V. Liseykina, F. Pegoraro, A. Macchi, Radiation-pressure-dominant acceleration: polarization and radiation reaction effects and energy increase in three-dimensional simulations. Phys. Rev. E **85**, 016407 (2012)
29. S.I. Abarzhi, Review of nonlinear dynamics of the unstable fluid interface: conservation laws and group theory. Phys. Scr. **2008**(T132), 014012 (2008)
30. S.I. Abarzhi, Nonlinear three-dimensional Rayleigh–Taylor instability. Phys. Rev. E **59**, 1729–1735 (1999)
31. Wikipedia (picture released as "public domain"). Byzantine marble pavement, Rome, 2007. Online; Accessed 28 June 2015
32. L. Michel, Symmetry defects and broken symmetry. Configurations hidden symmetry. Rev. Mod. Phys. **52**, 617–651 (1980)
33. D. Schattschneider, The plane symmetry groups: their recognition and notation. Am. Math. Mon. **85**(6), 439–450 (1978)
34. A. Macchi, A. Sgattoni, S. Sinigardi, M. Borghesi, M. Passoni, Advanced strategies for ion acceleration using high-power lasers. Plasma Phys. Control. Fusion **55**(12), 124020 (2013)
35. Y. Sentoku, K. Mima, S. Kojima, H. Ruhl, Magnetic instability by the relativistic laser pulses in overdense plasmas. Phys. Plasmas **7**(2), 689–695 (2000)
36. P. Mulser, D. Bauer, S. Hain, H. Ruhl, F. Cornolti, Present understanding of superintense laser-solid interaction. Laser Phys. **10**(1), 231–240 (2000)
37. F. Califano, D. Del Sarto, F. Pegoraro, Three-dimensional magnetic structures generated by the development of the filamentation (Weibel) instability in the relativistic regime. Phys. Rev. Lett. **96**, 105008 (2006)
38. B.F. Lasinski, A.B. Langdon, S.P. Hatchett, M.H. Key, M. Tabak, Particle-in-cell simulations of ultra intense laser pulses propagating through overdense plasma for fast-ignitor and radiography applications. Phys. Plasmas **6**(5), 2041–2047 (1999)

39. Y. Sentoku, K. Mima, Z.M. Sheng, P. Kaw, K. Nishihara, K. Nishikawa, Three-dimensional particle-in-cell simulations of energetic electron generation and transport with relativistic laser pulses in overdense plasmas. Phys. Rev. E **65**, 046408 (2002)
40. A. L'Huillier, P. Balcou, High-order harmonic generation in rare gases with a 1-ps 1053-nm laser. Phys. Rev. Lett. **70**, 774–777 (1993)
41. P.M. Paul, E.S. Toma, P. Breger, G. Mullot, F. Augé, P. Balcou, H.G. Muller, P. Agostini, Observation of a train of attosecond pulses from high harmonic generation. Science **292**(5522), 1689–1692 (2001)
42. P.B. Corkum, Plasma perspective on strong field multiphoton ionization. Phys. Rev. Lett. **71**, 1994–1997 (1993)
43. G. Sansone, E. Benedetti, F. Calegari, C. Vozzi, L. Avaldi, R. Flammini, L. Poletto, P. Villoresi, C. Altucci, R. Velotta, S. Stagira, S. De Silvestri, M. Nisoli, Isolated single-cycle attosecond pulses. Science **314**(5798), 443–446 (2006)
44. C. Ott, A. Kaldun, L. Argenti, P. Raith, K. Meyer, M. Laux, Y. Zhang, A. Blattermann, S. Hagstotz, T. Ding, R. Heck, J. Madronero, F. Martin, T. Pfeifer, Reconstruction and control of a time-dependent two-electron wave packet. Nature **516**(7531), 374–378 (2014)
45. F. Krausz, M. Ivanov, Attosecond physics. Rev. Mod. Phys. **81**, 163–234 (2009)
46. D. von der Linde, T. Engers, G. Jenke, P. Agostini, G. Grillon, E. Nibbering, A. Mysyrowicz, A. Antonetti, Generation of high-order harmonics from solid surfaces by intense femtosecond laser pulses. Phys. Rev. A **52**, R25–R27 (1995)
47. M. Yeung, B. Dromey, C. Rödel, J. Bierbach, M. Wünsche, G. Paulus, T. Hahn, D. Hemmers, C. Stelzmann, G. Pretzler, M. Zepf, Near-monochromatic high-harmonic radiation from rel ativistic laser-plasma interactions with blazed grating surfaces. New J. Phys. **15**(2), 025042 (2013)
48. F. Quéré, C. Thaury, P. Monot, S. Dobosz, P. Martin, J.-P. Geindre, P. Audebert, Coherent wake emission of high-order harmonics from overdense plasmas. Phys. Rev. Lett. **96**, 125004 (2006)
49. S.V. Bulanov, N.M. Naumova, F. Pegoraro, Interaction of an ultrashort, relativistically strong laser pulse with an overdense plasma. Phys. Plasmas **1**(3), 745–757 (1994)
50. R. Lichters, J. Meyer-ter Vehn, A. Pukhov, Short-pulse laser harmonics from oscillating plasma surfaces driven at relativistic intensity. Phys. Plasmas **3**(9), 3425–3437 (1996)
51. T. Baeva, S. Gordienko, A. Pukhov, Theory of high-order harmonic generation in relativistic laser interaction with overdense plasma. Phys. Rev. E **74**, 046404 (2006)
52. S. Gordienko, A. Pukhov, O. Shorokhov, T. Baeva, Relativistic doppler effect: universal spectra and zeptosecond pulses. Phys. Rev. Lett. **93**, 115002 (2004)
53. R.L. Sandberg, A. Paul, D.A. Raymondson, S. Hädrich, D.M. Gaudiosi, J. Holtsnider, R.I. Tobey, O. Cohen, M.M. Murnane, H.C. Kapteyn, C. Song, J. Miao, Y. Liu, F. Salmassi, Lensless diffractive imaging using tabletop coherent high-harmonic soft-X-ray beams. Phys. Rev. Lett. **99**, 098103 (2007)
54. J.L. Sanz-Vicario, H. Bachau, F. Martín, Time-dependent theoretical description of molecular autoionization produced by femtosecond xuv laser pulses. Phys. Rev. A **73**, 033410 (2006)
55. S.G. Rykovanov, M. Geissler, J. Meyer ter Vehn, G.D. Tsakiris, Intense single attosecond pulses from surface harmonics using the polarization gating technique. New J. Phys. **10**(2), 025025 (2008)
56. Z.-Y. Chen, X.-Y. Li, L.-M. Chen, Y.-T. Li, W.-J. Zhu, Intense isolated few-cycle attosecond xuv pulses from overdense plasmas driven by tailored laser pulses. Opt. Express **22**(12), 14803–14811 (2014)
57. X. Lavocat-Dubuis, J.-P. Matte, Numerical simulation of harmonic generation by relativistic laser interaction with a grating. Phys. Rev. E **80**, 055401 (2009)
58. X. Lavocat-Dubuis, J.-P. Matte, Numerical and theoretical study of the generation of extreme ultraviolet radiation by relativistic laser interaction with a grating. Phys. Plasmas **17**(9), 093105 (2010)
59. M. Yeung, M. Zepf, M. Geissler, B. Dromey, Angularly separated harmonic generation from intense laser interaction with blazed diffraction gratings. Opt. Lett. **36**(12), 2333–2335 (2011)

60. A. Aubry, D.Y. Lei, A.I. Fernández-Domínguez, Y. Sonnefraud, S.A. Maier, J.B. Pendry, Plasmonic light-harvesting devices over the whole visible spectrum. Nano Lett. **10**(7), 2574–2579 (2010)
61. A. Giugni, B. Torre, A. Toma, M. Francardi, M. Malerba, A. Alabastri, R. Proietti Zaccaria, M.I. Stockman, E. Di Fabrizio, Hot-electron nanoscopy using adiabatic compression of surface plasmons. Nat. Nanotechnol. **8**(11), 845–852 (2013)

Chapter 7
Conclusions and Perspectives

In this manuscript the role of plasmonic effects in several scenarios involving ultra-high intensity laser-matter interaction was investigated. Electron acceleration with relativistic surface plasmons was studied with an experimental campaign and a numerical simulation campaign. In addition, a theoretical model for the acceleration process was provided. Enhanced ion acceleration with foam-attached targets (which should enhanced the conversion of laser energy into bulk plasmons) was investigated numerically and experimentally. A model for laser-driven Rayleigh–Taylor instability in radiation pressure acceleration was proposed and numerical simulations were performed to support it. Plasmonic field enhancement plays a key role in this instability. Some preliminary results on enhanced high order harmonic generation with grating targets irradiated at their resonant angle for surface plasmon excitation were presented. Finally, the idea of porting some schemes from traditional plasmonics (essentially schemes for energy concentration) into the high field regime is discussed.

All the aforementioned topics hold promises for interesting further developments.

Since electron acceleration with relativistic surface plasmons seems to provide electron bunches suitable for a few applications and with properties not easily attainable with other sources, pursuing the development and optimization of this scheme could be a promising route. Optimization of this acceleration strategy may increase the total accelerated charge, hopefully up to ~ 1 nC. A proposal for a dedicated experimental campaign at Astra-Gemini facility (RAl, UK) and involving our group was recently submitted.

As far as ion acceleration with foam-attached targets is of concern, a few interesting routes can be pursued. Certainly the high energies obtained with a lengthening of the laser pulse deserves a more in-depth, dedicated, investigation. Moreover, since foam-attached targets are expected to be not dramatically sensitive to the pulse contrast, operating without the plasma mirror is certainly attractive (more energy on target could further enhance ion acceleration process).

Enhanced high order harmonic generation with grating targets irradiated at their resonance angle could be tested in a laboratory, since similar experiments have been performed in the past [1]. Moreover, in the numerical investigations, no attempts were performed to optimize harmonic generation process, for example with a parametric exploration of groove shape (e.g. testing blazed gratings like in [1] could be interesting).

Finally, the study of plasmonic waveguides or other schemes for energy concentration in the high field regime is certainly attractive. Up to now, only a very preliminary investigation was performed, while these schemes certainly deserve further investigations.

Reference

1. M. Yeung, B. Dromey, C. Rödel, J. Bierbach, M. Wünsche, G. Paulus, T. Hahn, D. Hemmers, C. Stelzmann, G. Pretzler, M. Zepf, Near-monochromatic high-harmonic radiation from relativistic laser-plasma interactions with blazed grating surfaces. New J. Phys. **15**(2), 025042 (2013)

Appendix A
Code Normalization

Everything in the code scales with ℓ_0 scale-length. m_e and q_e are, respectively, the mass and the charge of an electron, while c is the speed of light.

Normalization of Code Quantities

$$q = \tilde{q}\, q_e$$
$$m = \tilde{m}\, m_e$$
$$\mathbf{E} = \tilde{\mathbf{E}}\, \frac{m_e c^2}{q_e \ell_0}$$
$$\mathbf{B} = \tilde{\mathbf{B}}\, \frac{m_e c^2}{q_e \ell_0}$$
$$t = \tilde{t}\, \frac{\ell_0}{c}$$
$$l = \tilde{l}\, \ell_0$$
$$\mathbf{p} = \tilde{\mathbf{p}}\tilde{m}\, m_e c$$
$$\mathbf{v} = \tilde{\mathbf{v}}\, c$$
$$\mathbf{J} = \tilde{\mathbf{J}}\, \frac{m_e c^3 \pi}{q_e \ell_0}$$

Normalized Particle Equations

$$\begin{cases} \partial_{\tilde{t}} \tilde{p} = \dfrac{\tilde{q}}{\tilde{m}} \left(\tilde{\mathbf{E}} + \tilde{\mathbf{v}} \times \tilde{\mathbf{E}} \right) \\ \tilde{\mathbf{J}} = \tilde{q} \sum_n \dfrac{W_n}{\widetilde{\Delta V}} \tilde{\mathbf{v}}_n \end{cases}$$

Normalized Maxwell's Equations

$$\begin{cases} \partial_{\tilde{t}} \tilde{\mathbf{B}} = -\tilde{\nabla} \times \tilde{\mathbf{E}} \\ \partial_{\tilde{t}} \tilde{\mathbf{E}} = \tilde{\nabla} \times \tilde{\mathbf{B}} - 4\pi^2 \tilde{\mathbf{J}} \end{cases}$$

Appendix B
Particle-In-Cell algorithm

Code Units

Code units are normalized in such a way that $c = m_e = q_e = 1$, where c (2.99792×10^8 m/s) is the speed of light, q_e (1.60218×10^{-19} C) is the elementary charge and m_e (9.10938×10^{-31} kg) is the electron mass. Appendix A reports schematically the normalization conventions used throughout this document, unless otherwise stated. Normalized quantities are marked with a tilde (e.g. $\mathcal{E}$ is the electric field in SI units, while $\tilde{\mathcal{E}}$ is the electric field in normalized code units).

Boris Pusher for Particle Momentum

As far as Eq. 3.1 is of concern, in PIC method the distribution function is sampled with a family of macroparticles for each particle species. These macroparticles have definite momenta but they are extended in space (see Fig. B.1). This "diffusiveness" in space is beneficial to reduce numerical noise in the simulation results (a common issue in PIC codes).

Formally, the plasma distribution function is sampled as follows:

$$f(\mathbf{q}, \mathbf{p}, t) = f_0 \sum_{i=1}^{N} \delta(\mathbf{p} - \mathbf{p}_i(t)) S(\mathbf{q}, \mathbf{q}_i(t)) \tag{B.1}$$

where $\mathbf{q_i}(t)$ and $\mathbf{p}_i(t)$ are, respectively, position and momentum of the i-th macroparticle of the family. f_0 is simply a normalization constant, while $S(\mathbf{q}, \mathbf{q_i}(t))$ is the spatial shape function of a macroparticle. In 1D a typical choice is a triangular distribution centred at $\mathbf{q_i}(t)$ and with a total size of two Δx (Δx is the cell spacing). In 3D a natural extension is a shape function which is simply the product of three 1D shape functions. Using higher-order shape functions (which extends over more cells)

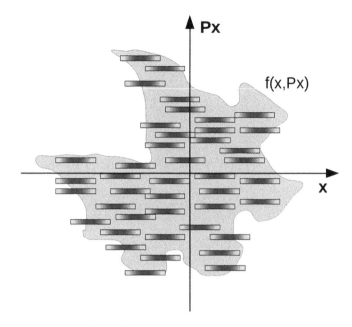

Fig. B.1 Sampling of the plasma distribution function with macroparticles. Each macroparticle has a definite momentum, but it is "diffused" in space.

generally results in a reduction of the numerical noise at the expense of a greater computational cost.

If we insert Eq. B.1 into Eq. 3.1, we get:

$$\sum_{i=1}^{N} \partial_t \left(\delta(\mathbf{p} - \mathbf{p}_i(t)) S(\mathbf{q}, \mathbf{q}_i(t)) \right) + \frac{\mathbf{p}}{m} \cdot \nabla_q \left(\delta(\mathbf{p} - \mathbf{p}_i(t)) S(\mathbf{q}, \mathbf{q}_i(t)) \right)$$
$$+ \mathbf{F}_L \cdot \nabla_p \left(\delta(\mathbf{p} - \mathbf{p}_i(t)) S(\mathbf{q}, \mathbf{q}_i(t)) \right) = 0 \qquad (B.2)$$

where the Lorentz force $\left(\mathbf{E} + \frac{1}{c} \mathbf{v} \times \mathbf{B} \right)$ was replaced with $\mathbf{F}_L$ and the velocity $\mathbf{v}$ with $\frac{\mathbf{p}}{m}$.

Taking into account the structure of the shape function, $S(\mathbf{q}, \mathbf{q}_i(t))$ can be rewritten as $S(\mathbf{q} - \mathbf{q}_i(t))$.

Proceeding with the calculation we get:

Appendix B: Particle-In-Cell algorithm

$$\sum_{i=1}^{N} -\delta(\mathbf{p}-\mathbf{p}_i(t))\dot{\mathbf{q}}_i(t)\cdot\nabla_q S(\mathbf{q}-\mathbf{q}_i(t)) - S(\mathbf{q}-\mathbf{q}_i(t))\dot{\mathbf{p}}\cdot\nabla_p\delta(\mathbf{p}-\mathbf{p}_i(t))+$$
$$\delta(\mathbf{p}-\mathbf{p}_i(t))\frac{\mathbf{p}}{m}\cdot\nabla_q S(\mathbf{q}-\mathbf{q}_i(t))+ \quad\quad (B.3)$$
$$S(\mathbf{q}-\mathbf{q}_i(t))\mathbf{F}_L(\mathbf{q},\mathbf{p}_i,t)\cdot\nabla_p\delta(\mathbf{p}-\mathbf{p}_i(t)) = 0$$

Since the terms with the δ functions are zero if $\mathbf{p}\neq\mathbf{p}_i$, we can perform the replacement:

$$\sum_{i=1}^{N} -\delta(\mathbf{p}-\mathbf{p}_i(t))\dot{\mathbf{q}}_i(t)\cdot\nabla_q S(\mathbf{q}-\mathbf{q}_i(t)) - S(\mathbf{q}-\mathbf{q}_i(t))\dot{\mathbf{p}}_i\cdot\nabla_p\delta(\mathbf{p}-\mathbf{p}_i(t))+$$
$$\delta(\mathbf{p}-\mathbf{p}_i(t))\frac{\mathbf{p}_i}{m}\cdot\nabla_q S(\mathbf{q}-\mathbf{q}_i(t))+ \quad\quad (B.4)$$
$$S(\mathbf{q}-\mathbf{q}_i(t))\mathbf{F}_L(\mathbf{q},\mathbf{p}_i,t)\cdot\nabla_p\delta(\mathbf{p}-\mathbf{p}_i(t)) = 0$$

Integrating this equation with respect to $d\mathbf{q}$ and $d\mathbf{p}$ will give the equations of motion for the macroparticles.

First, let's integrate over $d\mathbf{p}$ (two of the four previous terms disappear):

$$\sum_{i=1}^{N} -\dot{\mathbf{q}}_i(t)\cdot\nabla_q S(\mathbf{q}-\mathbf{q}_i(t)) + \frac{\mathbf{p}_i}{m}\cdot\nabla_q S(\mathbf{q}-\mathbf{q}_i(t)) = 0 \quad\quad (B.5)$$

Thus we get the simple solution:

$$\dot{\mathbf{q}}_i(t) = \frac{\mathbf{p}_i}{m}(t) \quad\quad (B.6)$$

Integration with respect to $d\mathbf{p}$ requires additional care. In fact, we should require the shape function to be normalized to 1 (a very reasonable assumption) and that $\int d\mathbf{q}\nabla_q S(\mathbf{q}-\mathbf{q_n}) = 0$.[1] With these assumptions we get:

$$\sum_{i=1}^{N}\left(-\dot{\mathbf{p}}_i + \int d\mathbf{q} S(\mathbf{q}-\mathbf{q_n})\mathbf{F}_L(\mathbf{q},\mathbf{p}_i,t)\right)\cdot\nabla_p\delta(\mathbf{p}-\mathbf{p}_i(t)) \quad\quad (B.7)$$

This leads to the simple solution:

$$\dot{\mathbf{p}}_i(t) = \int d\mathbf{q} S(\mathbf{q}-\mathbf{q_n})\mathbf{F}_L(\mathbf{q},\mathbf{p}_i,t) \quad\quad (B.8)$$

---

[1]This is automatically satisfied taking S to be the product of three triangular 1D shapes, one for each dimension.

Equations B.6 and B.8 are the equations of motion for the macroparticles. Each macroparticle moves as a particle of mass m under the effect of the Lorentz force averaged over its volume.

Boris Pusher for Second Order Accuracy in Time

The most widely used algorithm for Particle-In-Cell codes is the so-called *Boris pusher*, which ensures second order accuracy in time. Other higher order algorithms (e.g. Runge-Kutta4) may be useful in some specific applications. Higher order algorithms generally require more computational resources (more elementary operations and possibly more stored memory) per time-step, however they may allow for larger time-steps. Here only the classical Boris pusher will be presented, the reader interested in other higher order methods may be referred to [1] (also the following treatment of the Boris pusher algorithm is based on this reference).

The Boris Pusher technique is a leap-frog algorithm, which means that positions and momenta are known respectively at integer and half-integer times.

The equation for the evolution of the particle momentum from $t = n - 1/2$ to $t = n + 1/2$ reads as follows:

$$\mathbf{p}^{n+1/2} = \mathbf{p}^{n-1/2} + \Delta t \frac{q}{m} \left[\mathbf{E}^n + \mathbf{v}^n \times \mathbf{B}^n \right] \quad (B.9)$$

where the superscript a^n indicates that the quantity is considered at time $t = n$. $\mathbf{E}^n$ and $\mathbf{B}^n$ are, respectively, the electric field and the magnetic field at time $t = n$ and interpolated at the center of the macro-particle $\mathbf{r}^n$ at $t = n$. Since $\mathbf{E}$ and $\mathbf{B}$ are leap-frogged in time, this requires that $\mathbf{B}$ is advanced in two separate half-steps.

Equation B.9 requires $\mathbf{v}^n = \mathbf{p}^n / \gamma^n$. However $\mathbf{p}$ is known only at half-integer times. Consequently we may approximate $\mathbf{p}^n$ as[2]

$$\mathbf{p}^n = \frac{\mathbf{p}^{n+2} - \mathbf{p}^{n-1/2}}{2}. \quad (B.10)$$

We end up with the following implicit equation:

$$\mathbf{p}^n = \mathbf{p}^{n-1/2} + \frac{\Delta t}{2} \frac{q}{m} \left[\mathbf{E}^n + \frac{\mathbf{p}^n}{\gamma^n} \times \mathbf{B}^n \right] \quad (B.11)$$

We define the following quantities:

---

[2] This approximation is third-order accurate, as noted in [1].

Appendix B: Particle-In-Cell algorithm

$$\beta = \frac{q}{m}\Delta t/2 \tag{B.12}$$

$$\mathbf{b} = \beta \mathbf{B}^n/\gamma^n \tag{B.13}$$

$$\tilde{\mathbf{p}} = \mathbf{p}^{n-1/2} + \beta \mathbf{E}^n \tag{B.14}$$

Using these definitions, Eq. B.11 can be rewritten as follows:

$$\mathbf{p}^n = \tilde{\mathbf{p}} + \mathbf{p}^n \times \mathbf{b} \tag{B.15}$$

$$\gamma^n = \sqrt{1 + \tilde{\mathbf{p}} \cdot \tilde{\mathbf{p}}} \tag{B.16}$$

$$(\mathbf{p}^n - \mathbf{p}^n \times \mathbf{b}) \times \mathbf{b} = \tilde{\mathbf{p}} \times \mathbf{b} \tag{B.17}$$

$$\mathbf{p}^n = \frac{1}{1+b^2}\left[\tilde{\mathbf{p}} + \tilde{\mathbf{p}} \times \mathbf{b} + \mathbf{b}(\tilde{\mathbf{p}} \times \mathbf{b})\right] \tag{B.18}$$

The particle momentum at time $t = n + 1/2$ is finally calculated as:

$$\mathbf{p}^{n+1/2} = 2\mathbf{p}^n - \mathbf{p}^{n-1/2} \tag{B.19}$$

Particle positions are simply advanced by Δt from $t = n$ to $t = n + 1$ using their momentum, known at time $t = n + 1/2$:

$$\mathbf{r}^{n+1} = \mathbf{r}^n + \Delta t \mathbf{v}^{n+1/2} \tag{B.20}$$

FDTD Maxwell Solver with Yee-Lattice

In a PIC code, a Maxwell solver is needed in order to integrate in time the electromagnetic field equations. Several options exist, tailored at different computational needs. Here, a popular simple Maxwell solver -the second order FDTD solver on a yee lattice- is described, highlighting its features and limitations. This solver was first described in [2] and it is based on a leap-frog scheme in time and on a spatial discretization on a staggered grid (the idea of using a staggered grid to improve the accuracy of numerical solvers for hydrodynamics dates back to 1950, see [3]).

B field is shifted in time by a $\Delta t/2$ with respect to **E** field (i.e. time evolution of **E** field from step n to step $n + 1$ is calculated using **B** field at time step $n + 1/2$). In order to allow for centred spatial differentiation, fields are arranged on a 3D grid as shown in Fig. B.2 (along $\hat{x}$ axis, i is an integer grid point, while $i + 1/2$ is an half-integer grid point).

In conclusion, discretized Maxwell equations in this scheme are:

Fig. B.2 Staggered grid for Yee FDTD Maxwell solver

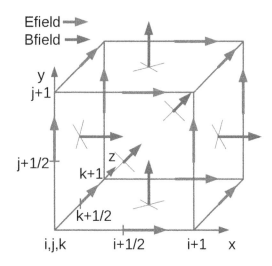

$$
\begin{cases}
\dfrac{E_x^{n+1}(i+1/2,j,k)-E_x^n(i+1/2,j,k)}{\Delta t} = \\
\quad +\left(\dfrac{B_z^{n+1/2}(i+1/2,j+1/2,k)-B_z^{n+1/2}(i+1/2,j-1/2,k)}{\Delta y}\right. \\
\quad \left. -\dfrac{B_y^{n+1/2}(i+1/2,j,k+1/2)-B_y^{n+1/2}(i+1/2,j,k-1/2)}{\Delta z}\right)+4\pi^2 J_x^{n+1/2}(i+1/2,j,k) \\[6pt]
\dfrac{E_y^{n+1}(i,j+1/2,k)-E_y^n(i,j+1/2,k)}{\Delta t} = \\
\quad +\left(\dfrac{B_x^{n+1/2}(i,j+1/2,k+1/2)-B_x^{n+1/2}(i,j+1/2,k-1/2)}{\Delta z}\right. \\
\quad \left. -\dfrac{B_z^{n+1/2}(i+1/2,j+1/2,k)-B_z^{n+1/2}(i-1/2,j+1/2,k)}{\Delta x}\right)+4\pi^2 J_y^{n+1/2}(i,j+1/2,k) \\[6pt]
\dfrac{E_z^{n+1}(i,j,k+1/2)-E_z^n(i,j,k+1/2)}{\Delta t} = \\
\quad +\left(\dfrac{B_y^{n+1/2}(i+1/2,j+1/2,k+1/2)-B_y^{n+1/2}(i-1/2,j+1/2,k+1/2)}{\Delta x}\right. \\
\quad \left. -\dfrac{B_x^{n+1/2}(i,j+1/2,k+1/2)-B_x^{n+1/2}(i,j-1/2,k+1/2)}{\Delta y}\right)+4\pi^2 J_z^{n+1/2}(i,j,k+1/2) \\[6pt]
\dfrac{B_x^{n+1/2}(i,j+1/2,k+1/2)-B_x^{n-1/2}(i,j+1/2,k+1/2)}{\Delta t} = \\
\quad -\left(\dfrac{E_z^n(i,j+1,k+1/2)-E_z^n(i,j,k+1/2)}{\Delta y}-\dfrac{E_y^n(i,j+1/2,k+1)-E_y^n(i,j+1/2,k)}{\Delta z}\right) \\[6pt]
\dfrac{B_y^{n+1/2}(i+1/2,j,k+1/2)-B_y^{n-1/2}(i+1/2,j,k+1/2)}{\Delta t} = \\
\quad -\left(\dfrac{E_x^n(i+1/2,j,k+1)-E_x^n(i+1/2,j,k)}{\Delta z}-\dfrac{E_z^n(i+1,j,k+1/2)-E_z^n(i,j,k+1/2)}{\Delta x}\right) \\[6pt]
\dfrac{B_z^{n+1/2}(i+1/2,j+1/2,k)-B_z^{n-1/2}(i+1/2,j+1/2,k)}{\Delta t} = \\
\quad -\left(\dfrac{E_y^n(i+1,j+1/2,k)-E_y^n(i,j+1/2,k)}{\Delta x}-\dfrac{E_x^n(i+1/2,j+1,k)-E_x^n(i+1/2,j,k)}{\Delta y}\right)
\end{cases}
$$

(B.21)

References

1. P. Londrillo, A. Sgattoni, and F. Rossi. The aladyn code technical report. Technical report, University of Bologna, INAF-Bologna and INFN-Bologna, 2011
2. K. Yee, Numerical solution of initial boundary value problems involving Maxwell's equations in isotropic media. Antennas and Propagation, IEEE Transactions on **14**(3), 302–307 (May 1966)
3. J. VonNeumann, R.D. Richtmyer, A method for the numerical calculation of hydrodynamic shocks. Journal of Applied Physics **21**(3), 232–237 (1950)

Curriculum Vitae

Luca Fedeli

Birth: 1988, Gallarate (Italy)
Current position: (as of May 2016) post-doc at Politecnico di Milano, Italy
Research interests: Massively parallel Particle-In-Cell simulations (also code development). Laser-plasma interaction at relativistic intensities. Laser-driven sources (ion and electron acceleration with laser-produced plasmas, high order harmonic generation). Instabilities in plasma astrophysics.
Other interests: I've been practising martial arts for three years (Jeet Kune Do). I'm interested in invertebrates, fractals and snorkeling. I enjoy reading and travelling.

Education

- **PhD in Physics**

    Dates 11/2012–12/2015
    Institution University of Pisa (Italy)

- **Master's degree in Physics**

    Dates 11/2010–10/2012
    Institution University of Milano-Bicocca (Italy)
    Final Mark 110/110 cum laude

- **Bachelor's degree in Physics**
  Dates 08/2007-10/2010
  Institution University of Milano-Bicocca (Italy)
  Final Mark 110/110 cum laude

Publications in Peer Reviewed Journals

- L. Fedeli, A. Sgattoni, G. Cantono, D. Garzella, F. Réau I. Prencipe, M. Passoni, M. Raynaud, M. Květoň, J. Proska, A.Macchi, T. Ceccotti. Electron acceleration by relativistic surface plasmons in laser-grating interaction. *Physical Review Letters* 116, 015001, 2016.
- M. Passoni, A. Sgattoni, I. Prencipe, L. Fedeli, D. Dellasega, L. Cialfi, I.W. Choi, I.J. Kim, K.F. Kakolee, K.A. Janulewicz, H.W. Lee, J.H. Sung, S.K. Lee, C.H. Nam. Toward flexible laser–driven ion beams: nanostructured double–layer targets. *Phys. Rev. Accel. Beams* 19, 061301 (2016)
- A. Grassi, L. Fedeli, A. Sgattoni, A. Macchi. Vlasov simulation of laser-driven collisionless shock/soliton acceleration and ion turbulence. *Plasma Physics and Controlled Fusion* 58, 034021 (2016).
- I. Prencipe, A. Sgattoni, D. Dellasega, L. Fedeli, L. Cialfi, I.W. Choi, I.J. Kim, K. Janulewicz, K. Kakolee, H.W. Lee, J.H. Sung, H. Jae, S.K. Lee, C.H. Nam, M. Passoni. Development of foam-based layered targets for laser-driven ion beam production. *Plasma Physics and Controlled Fusion*, 58, 034019 (2016).
- A. Sgattoni, L. Fedeli, G. Cantono, T. Ceccotti, A. Macchi. High field plasmonics and laser-plasma acceleration in solid targets. *Plasma Physics and Controlled Fusion*, 58, 014004 (2016).
- M. D'Angelo, L. Fedeli, A. Sgattoni, F. Pegoraro, and A. Macchi. Particle acceleration and radiation friction effects in the filamentation instability of pair plasmas. *Monthly Notices of the Royal Astronomical Society*, 451(4):3460–3467, 2015.
- A. Sgattoni, S. Sinigardi, L. Fedeli, F. Pegoraro, and A. Macchi. Laser-driven Rayleigh-Taylor instability: Plasmonic effects and three-dimensional structures. *Phys. Rev. E*, 91:013106, Jan 2015.
- A. Grassi, L. Fedeli, A. Macchi, S.V. Bulanov, and F. Pegoraro. Phase space dynamics after the breaking of a relativistic langmuir wave in a thermal plasma. *The European Physical Journal D*, 68(6), 2014.

These publications are based on work carried out before my PhD:

- L. Antonelli, P. Forestier-Colleoni, G. Folpini, R. Bouillaud, A. Faenov, <u>L. Fedeli</u>, C. Fourment, L. Giuffrida, S. Hulin, S.A. Pikuz Jr., J.J. Santos, L. Volpe, D. Batani. Measurement of reflectivity of spherically bent crystals using $K\alpha$ signal from hot electrons produced by laser-matter interaction. *Review of Scientific Instruments*, 86.7, 2015
- A. Morace, <u>L. Fedeli</u>, D. Batani, S. Baton, F. N. Beg, S. Hulin, L. C. Jarrott, A. Margarit, M. Nakai, M. Nakatsutsumi, P. Nicolai, N. Piovella, M. S. Wei, X. Vaisseau, L. Volpe, and J. J. Santos. Development of x-ray radiography for high energy density physics.*Physics of Plasmas (1994-present)*, 21(10):–, 2014.

Conference Proceedings

- <u>L. Fedeli</u>, A. Sgattoni, G. Cantono, I. Prencipe, M. Passoni, O. Klimo, J. Proska, A. Macchi, and T. Ceccotti. Enhanced electron acceleration via ultra-intense laser interaction with structured targets. *Proc. SPIE*, 9514:95140R–95140R–8, 2015.

Publications in Preparation

- <u>L. Fedeli</u>, A. Sgattoni, A. Macchi. Relativistic surface-plasmon enhanced harmonic generation from grating targets
- S. Kar, F. Hanton, D. Gwynne, H. Ahmed, <u>L. Fedeli</u>, A. Sgattoni, A. Macchi, D. Doria, M. Cerchez, A. Alejo, J. Fernandez, P. McKenna, D. Neely, J.A. Ruiz, O. Willi, M. Zepf and M. Borghesi. Narrow energy band, narrow divergence ion beams from laser driven ultrathin foils.

White Papers

- A. Sgattoni, <u>L. Fedeli</u>, S. Sinigardi, A. Marocchino, A. Macchi, V. Weinberg, and A. Karmakar. Optimising piccante – an open source particle-in-cell code for advanced simulations on tier-0 systems. Technical report, PRACE white papers, 2015. Online; accessed 23-May-2015.

Participation to Experiments

- 12/07/15–24/07/15 GIST, Gwangju (Republic of Korea)
- 28/10/14–12/11/14 GIST, Gwangju (Republic of Korea)

- 08/09/14–26/09/14 + 05/10/14–18/10/14 CEA-Saclay, Gif-sur-Yvette (Paris, France)
- 30/06/14–11/07/14 CEA-Saclay, Gif-sur-Yvette (Paris, France)
- 23/02/14–29/03/14 Rutherford Appleton Laboratory, Harwell (Oxford, UK).

Talks and Seminars

- Talk, 28 September 2015, FISMAT 2015 (Italian National Conference on Condensed Matter Physics), Palermo, Italy. *Plasmonics effects in relativisitic High Field regime: from electron acceleration to High Harmonics Generation.*
- Invited seminary, 27 July 2015, Osaka University, Japan. *Electron acceleration with laser-induced relativistic surface plasmons*
- Talk, Workshop on Ultraintense femtosecond laser-foil interaction, laser-plasma processes, fast electron transport and light ion acceleration, 21 May 2015, CNR-Pisa, Italy. *High–field relativistic plasmonics for laser–driven sources*
- Talk, SPIE Optics+Optoelectronics, 14 April 2015, Prague, Czech Republic. *Laser-driven proton and electron acceleration in high-field plasmonic regime*

Poster Contributions

- *Effects of Local Field Modulation on the Laser-Driven Rayleigh-Taylor Instability, plasmonic effects and 3D structures*, 16–20 November 2015, Savannah, Georgia (USA). [Presented by Prof F.Pegoraro (Univ. of Pisa)]
- *Simulation of laser-grating interactions for generation of high harmonics and extreme intensities* presented at Novel Light Sources from Laser-Plasma Interactions 2015, 20–24 April 2015, Dresden, Germany
- *Shock Propagation in Fast Ignition Cone Targets* presented at ELI Beamlines Summer School 2013, 23–28 June 2013, Prague, Czech Republic and at 2013 High-Energy-Density-Physics Summer School, 14th–20th July 2013, Columbus, Ohio (USA).

Attended Schools and Courses

- Introduction to the FERMI Blue Gene/Q, for users and developers, 18th March 2013, Casalecchio di Reno, Italy
- 2013 High-Energy-Density-Physics Summer School, 14th–20th July 2013, Columbus, Ohio (USA).
- ELI Beamlines Summer School 2013, 23–28 June 2013, Prague, Czech Republic.

CPSIA information can be obtained
at www.ICGtesting.com
Printed in the USA
LVHW02*1331110318
569449LV00002B/432/P